NATURAL HISTORY OF ANIMALS.

CONTAINING

BRIEF DESCRIPTIONS OF THE ANIMALS FIGURED ON TENNEY'S NATURAL HISTORY TABLETS, BUT COMPLETE WITHOUT THE TABLETS.

BY

SANBORN TENNEY AND ABBY A. TENNEY.

ILLUSTRATED WITH FIVE HUNDRED WOOD ENGRAVINGS,
CHIEFLY OF NORTH AMERICAN ANIMALS.

SEVENTH EDITI

NEW YORK:
SCRIBNER, ARMSTRONG & CO.,
654 BROADWAY.
1874.

JOHN F. TROW & SON,
PRINTERS AND BOOKBINDERS,
205–213 *East 12th St.*,
NEW YORK.

THIS BRIEF ACCOUNT

OF

THE ANIMAL KINGDOM

IS AFFECTIONATELY DEDICATED

TO THE YOUNG.

PREFACE.

THIS little volume contains a brief account of the Animal Kingdom, and it is hoped that it may aid Parents and Teachers in interesting the young in the delightful and important study of Natural History. As indicated on the title-page, it serves the purpose of a key to the Natural History Tablets, but is also complete in itself without the Tablets.

It is proper to add, that the engravings are the same, with few exceptions, as those in Tenney's "Manual of Zoölogy," and that those of the Mammals are mainly from Schinz, Audubon & Bachman, and Richardson; of the Birds, mainly from Audubon and Wilson; of the Reptiles and Batrachians, mainly from Holbrook; of the Fishes, from Storer, Holbrook, DeKay, and from nature; of the Insects, from Harris, Emmons, Say, Sanborn, and from nature; of the Crustaceans, mainly from nature and Reports; of the Mollusks, from Binney, Woodward, Gould, Lea, Conrad, and from nature; of the Echinoderms, from nature, Agassiz, and Müller; of the Acalephs, from Agassiz; of the Polyps, from Dana, Milne-Edwards, Verrill, and from nature; and of the Protozoans, mainly from Ehrenberg and Huxley.

Both this volume, and the "Elements of Zoölogy" already announced by Messrs. Scribner & Co., and which will combine the study of the Anatomy and Physiology of Animals with that of Descriptive Zoology, are intended to precede the Manual mentioned above.

VASSAR COLLEGE, POUGHKEEPSIE, N. Y.,
August, 1866.

CONTENTS.

TENNEY'S NATURAL HISTORY TABLETS.

THE Natural History Tablets referred to on the title-page are five in number.

No. 1. MAMMALS.

No. 2. BIRDS.

No. 3. REPTILES AND FISHES.

No. 4. INSECTS, CRUSTACEANS, AND WORMS.

No. 5. MOLLUSKS OR SHELL-FISH, SEA-URCHINS, STAR-FISHES, JELLY-FISHES, SEA-ANEMONES, AND CORALS.

These Tablets are adapted for use in Schools and in the Family, where it is believed that they may be efficient aids in interesting and instructing the young in the important subjects which they illustrate. Both the popular and the scientific names are generally given under each animal figured. The page where the animal is described in this book may be readily found by reference to the Index.

It is hoped that Teachers who desire to give "Object Lessons" in Natural History will find in the "Tablets" and "Natural History of Animals" such helps as they most need.

NATURAL HISTORY OF ANIMALS.

A GENERAL IDEA OF ANIMALS.

ANIMALS are living beings which feed upon plants, — or, in many cases, upon animals whose food is plants, — and which have the sense of feeling and the power of motion. The kinds of animals are very numerous, — more numerous than the kinds of trees in the forest and the flowers of the meadows and fields; and they are of all sizes, from those so minute that thousands can sport in a drop of water, to those of large dimensions, like the Horse and the Ox, the Elephant and the Whale; and their forms are as various as their sizes and kinds. But the name *Animal* is given to them all, whatever their size or form, and whether they swim, creep, fly, walk, or run.

Animals are most interesting objects for study, and the child as well as the man is delighted with learning their forms, structure, color, habits, and names, and soon becomes as eager as a naturalist to find a new Bird or a new Butterfly.

Some kinds of animals, as Man, Cattle, Deer, Sheep, Beasts of Prey, Birds, Turtles, Lizards, Snakes, Frogs, and Fishes, have a backbone containing a spinal cord, which is enlarged at the forward end into an organ

Fig. 1. — Deer — American Elk.

Fig. 2 — Bird — Duck.

called the brain; and as the backbone is made up of parts called vertebræ, these animals have been named VERTEBRATES. See Figures 1–6.

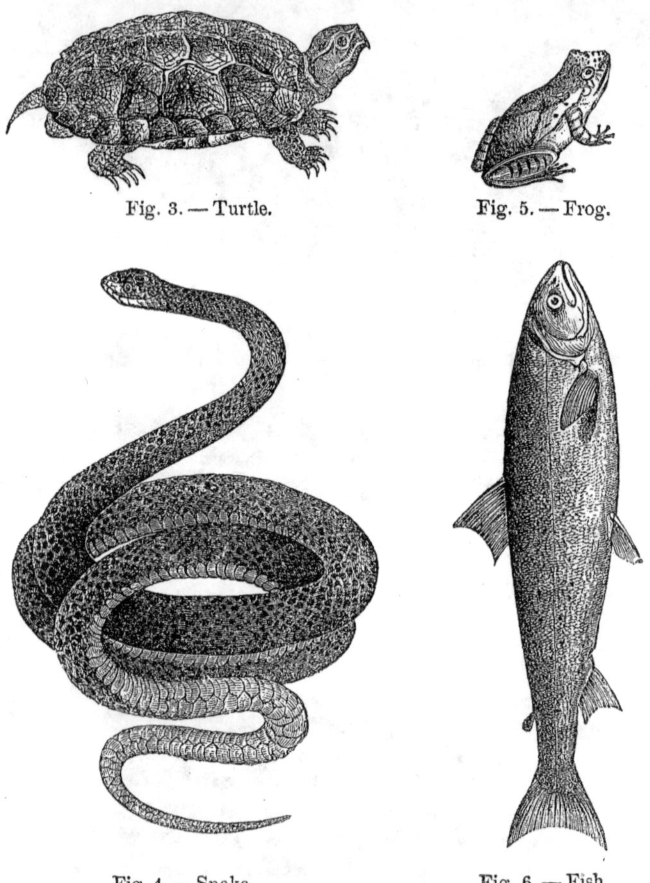

Fig. 3. — Turtle.

Fig. 5. — Frog.

Fig. 4. — Snake.

Fig. 6. — Fish.

Other kinds of animals, as Bees, Butterflies, Flies, and all other Insects, together with Crabs, Lobsters,

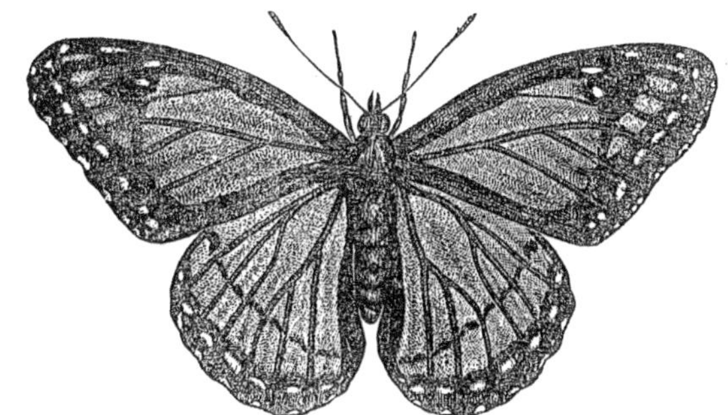

Fig. 7. — Butterfly.

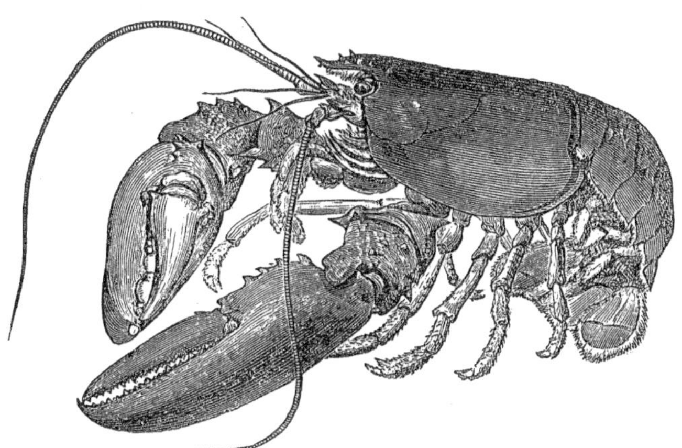

Fig. 8. — Lobster.

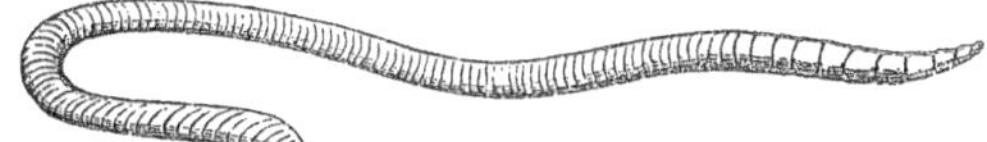

Fig. 9. — Worm.

Shrimps, and Worms, are made up of a series of rings, or joints, and hence are called ARTICULATES, from a word which means jointed. See Figures 7–9.

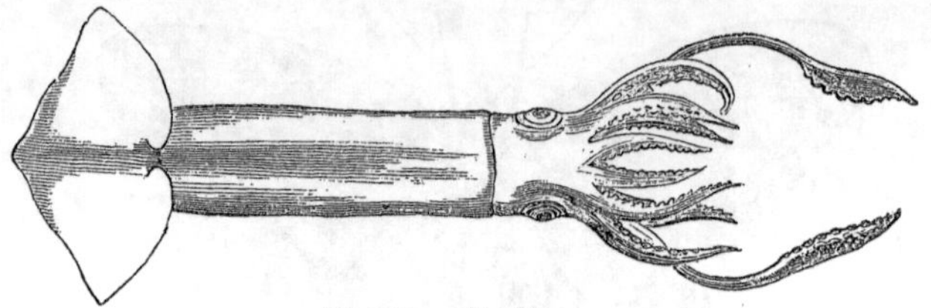

Fig. 10.— Squid.

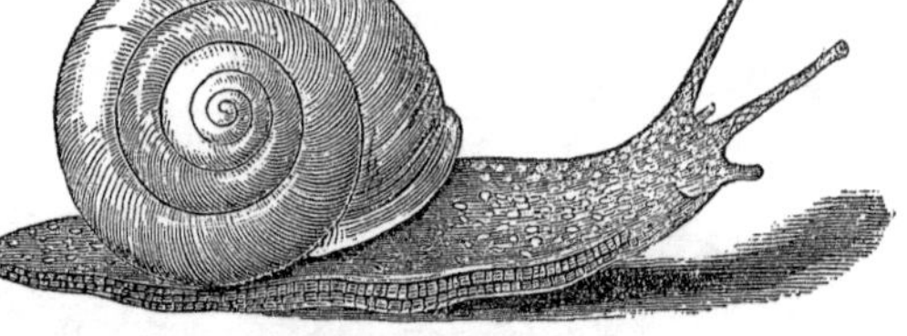

Fig. 11.— Land Snail.

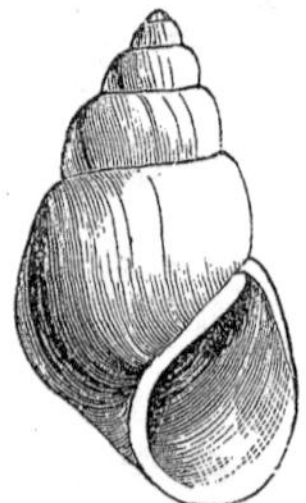

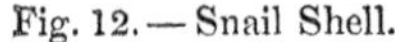

Fig. 12.— Snail Shell.

Fig. 13.— Fresh-Water Mussel.

Other kinds of animals, as Squids, Snails, Mussels, Clams, and Oysters, have neither a backbone nor a jointed body; but the whole body is soft, sometimes with a shell outside, and sometimes without a shell, and they are called MOLLUSKS, from a word which means soft. See Figures 10–13.

Still other kinds of animals, as Sea-Urchins, Sea-Stars, Jelly-Fishes, Sea-Anemones, and Coral-Polyps,

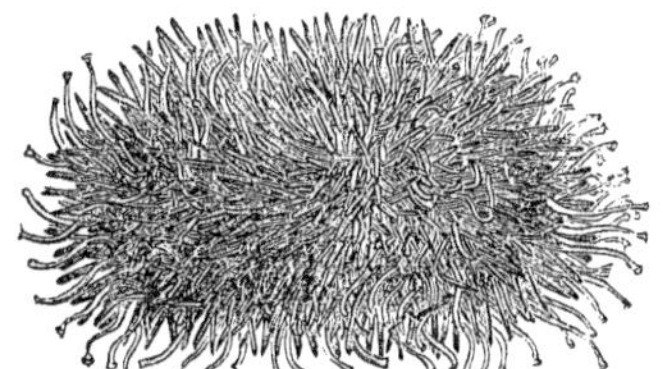

Fig. 14. — Sea-Urchin.

Fig. 15. — Sea-Star or Star-Fish.

are star-shaped, or flower-shaped, their parts radiating from a common centre or axis, and hence these animals are called RADIATES. See Figures 14–19.

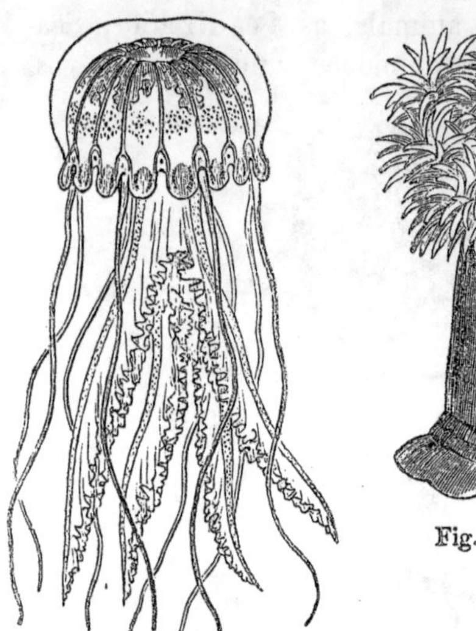

Fig. 16. — Jelly-Fish.

Fig. 17. — Sea-Anemone.

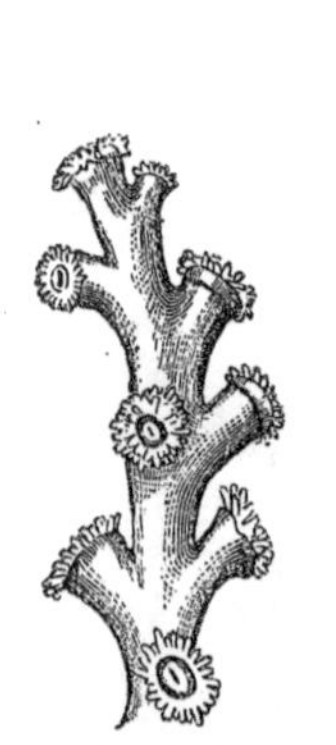

Fig. 18. — Coral-Polyps.

Fig. 19. — Coral-Polyps.

VERTEBRATES, OR BACKBONED ANIMALS.

THE Vertebrates, as stated on the first page, have a backbone made of parts, each one of which is called a vertebra. This backbone is the most important portion of a bony framework called a skeleton. Upon this skeleton is placed the flesh, and outside of the whole is the skin, which is naked, or covered with hair, fur, feathers, or scales, according to the kind of animal. Within the head is a wonderful organ called the brain, which has a branch called the spinal cord, extending the whole length of the body, and contained in a tube formed in the upper part of the backbone. From the spinal cord there are little branches, called nerves, which reach to all parts of the body. The brain, spinal cord, and nerves are called the Nervous System, which is much the same in its general

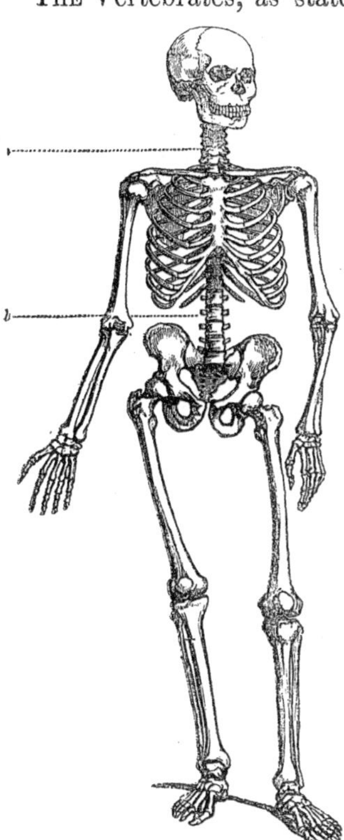

Fig. 20. — Skeleton of the highest Vertebrate — Man.

b, *b*, backbone.

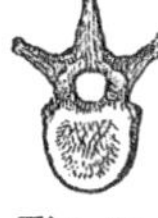

A single vertebra, the round white space showing the place of the spinal cord.

Fig. 21.

character in all vertebrates. This system as it appears in Man, the highest vertebrate, is shown in Fig. 22. Besides the brain and spinal cord, the skeleton protects the organs for breathing, digestion, and other organs peculiar to animals.

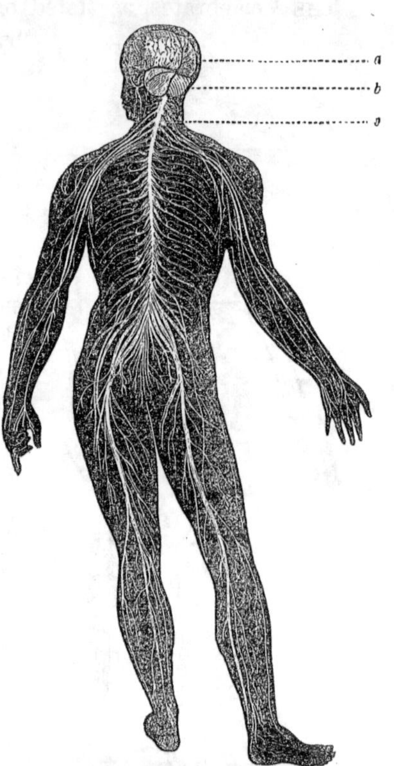

Fig. 22.—Nervous System of the highest Vertebrate—Man.

a, principal brain, called the hemispheres; *b*, smaller brain; *c*, spinal cord giving off its branches of nerves.

As the brain and spinal cord are alike in their position and general outlines in all vertebrates, only differing in extent and in degrees of perfection, so also are the skeletons of all vertebrates alike in their principal features. The back-bone of one, in its position and general outlines, corresponds to that of all the others; so with the head, the neck, and limbs. The arm of Man, the arm of a Monkey, the wing of a Bat, the leg of a Mole, the leg of a Dog, the paddle of a Seal, the leg of a Sheep, the paddle of a Whale, the wing of a Bird, the leg of a Turtle, and the fin of a

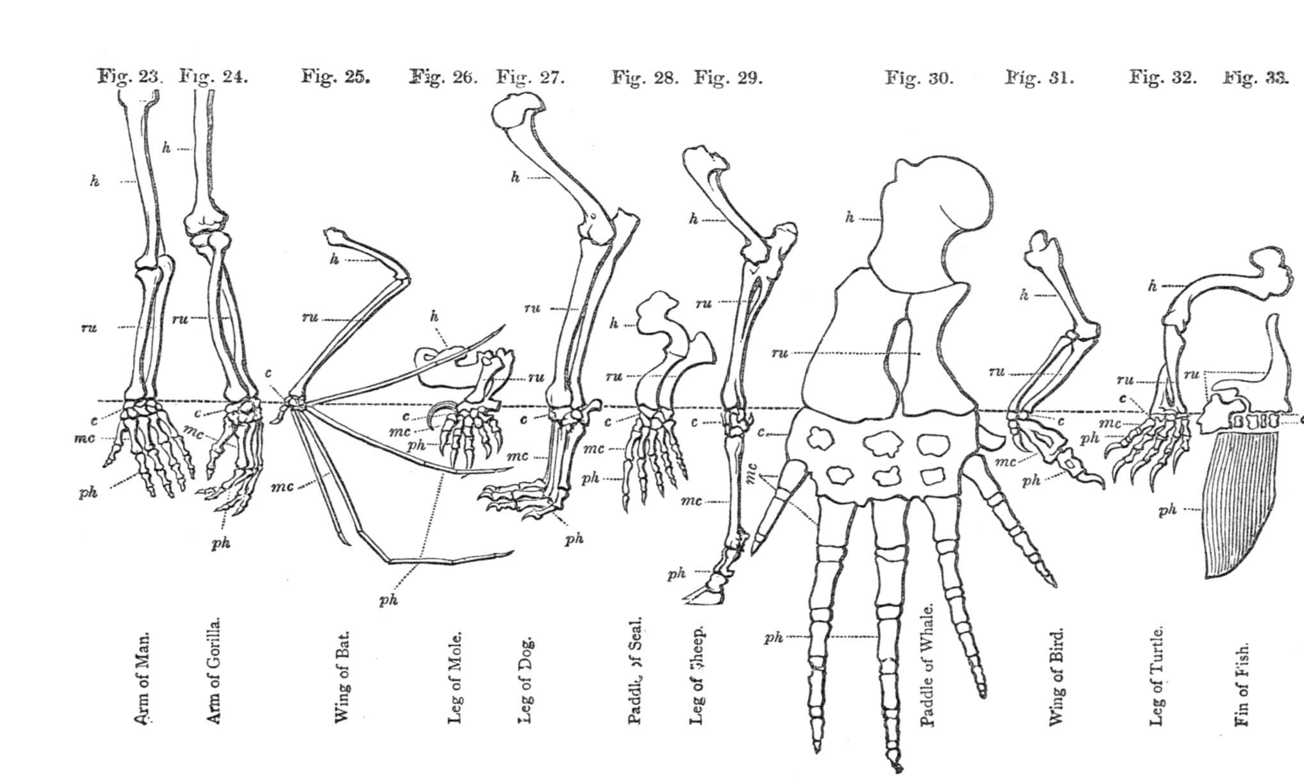
Fig. 23.
Fig. 24.
Fig. 25.
Fig. 26.
Fig. 27.
Fig. 28.
Fig. 29.
Fig. 30.
Fig. 31.
Fig. 32.
Fig. 33.
h
ru
c
mc
ph
Arm of Man.
Arm of Gorilla.
Wing of Bat.
Leg of Mole.
Leg of Dog.
Paddle of Seal.
Leg of Sheep.
Paddle of Whale.
Wing of Bird.
Leg of Turtle.
Fin of Fish.

Fish, correspond to one another in their most important features, each being modified according to the use for which it was made. This is quite plainly seen by studying Figs. 23–33, and observing that the corresponding parts are marked with the same letter.

The Vertebrates are divided into Mammals, Birds, Reptiles, Batrachians, and Fishes.

MAMMALS.

The Mammals are vertebrates which bring forth living young and nourish them with milk. Man, Monkeys, Beasts of Prey, Hoofed Animals, Whales, Bats, Moles, Squirrels and Rats, Sloths, Kangaroos and Opossums, and Duckbills, come under this head. They all breathe air by means of organs called lungs, have warm blood which is sent throughout the body by means of a heart constructed like that of Man, and the neck has only seven vertebræ.

MAN.

Man is at the head of the Animal Kingdom. He is the only animal to whom the upright position is natural; the only one which has a perfect hand; the only one whose forward extremities — arms and hands — are not used for locomotion; the only one that laughs; the only one that speaks a language; and his brain is larger than that of any other animal,* and he can live in all countries. But Man is also far more than an animal. He has a Mind and a Soul. He can learn much about the things which God has made, and understand the Bible which He has given.

* The brain of the Elephant and of the Whale is said to be larger than that of Man.

MONKEYS OR QUADRUMANA.

Monkeys are animals whose four feet are hand-like, and hence their scientific name, Quadrumana, which means *four-handed.* But though these hands are well

Fig. 34. — Chimpanzee.

adapted for grasping and climbing, they are much inferior to the perfect hand of Man. Some kinds can stand upright, but not firmly, for the soles of their feet nearly face each other, and cannot be brought flat to the ground like the foot of Man. About eighty kinds of Monkeys live in the forests of the warm parts of Asia and Africa, and even more kinds in South America. Those of Africa and Asia have thirty-two teeth, their nostrils near together, and their tail, even when present, is not capable

Fig. 35. — Orang-Outang.

of grasping objects. Most of the Monkeys of America have thirty-six teeth, the nostrils far apart, and many of them have the tail capable of grasping objects, and thus of being used in climbing and in picking up objects which cannot be reached by the hand. Monkeys live mainly on the trees, and feed upon fruits, nuts, eggs, and insects. They are selfish, mischievous, and thievish.

The Chimpanzee of Western Africa is one of the monkeys which has no tail, and is called an Ape, and, of all its tribe is thought to be the most like Man; but the great African Ape, called the Gorilla, is a larger species. Although when in an upright position the Chimpanzee somewhat resembles a human being, its long muzzle and other characters separate it widely even from the lowest tribes of the human family. The Orang-Outang is an ape which inhabits Borneo, and much resembles the Chimpanzee. Each of these is about as tall as a man. The Kahau of India is about the size of a large dog, and is named from its peculiar cry. The Baboons, often called Dog-headed Monkeys and Mandrills, have a very long muzzle, like that of a dog, as shown

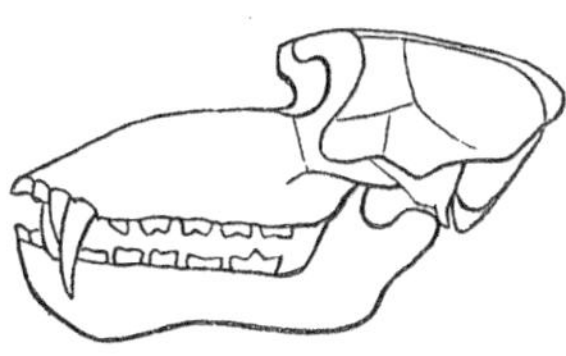

Fig. 36.— Skull of Baboon.

Fig. 37. — Kahau.

Fig. 38. — Spider Monkey.

by Fig. 36. They are common in Africa, and some of

them are very large and ferocious, and in appearance are the ugliest of all the Monkeys. The Spider-Monkey of South America is so called from its sprawling legs. Its long tail is of great aid in climbing. The Marmosets of Brazil are very small and curious monkeys, with long,

Fig. 39. — Marmoset.

Fig. 40. — Lemur.

soft, and beautifully colored fur. The Lemurs, or Makis, are pretty monkey-like animals, which live in Madagascar. The tail is quite bushy, and in many respects they much resemble common four-footed animals.

Fig. 41. — Aye-Aye.

The Aye-Aye is a curious monkey-like animal, about as large as a cat, which lives in Madagascar. Its incisor teeth are like those of the Rodents; its middle finger is exceedingly elongated and slender; and its tail bushy.

Some kinds of Monkeys imitate the actions of men,

and their efforts of this sort are often exceedingly ludicrous. In imitation of its master an ape has sat at table, using knife and fork, and drinking wine. It is stated that an ape owned by a French priest once followed him to church and hid upon the sounding-board, — a fixture over the pulpit, — and, when the sermon was going on, advanced to the edge of the board, and, observing the actions of the preacher, began to perform also, and his imitations were so perfect that the whole congregation were unable to suppress their laughter. The priest was shocked and indignant at such levity, and commenced to give his audience severe reproofs; but seeing all his efforts failing, his action became more violent and his voice louder; but his violent gestures were taken up by the ape with no less animation than that shown by his master, and at this apparent competition of the two the people burst into laughter louder than before.

FLESH-EATERS, OR CARNIVORES.

These animals have their teeth and claws very sharp, and they capture and devour other animals for food. Their back teeth, or molars, have sharp edges, and those in the two jaws shut by each other like the blades of scissors, and thus cut the flesh into pieces fit for swallowing. Cats, Hyenas, Dogs, Civets, Weasels, Bears, and Seals are the principal Carnivores.

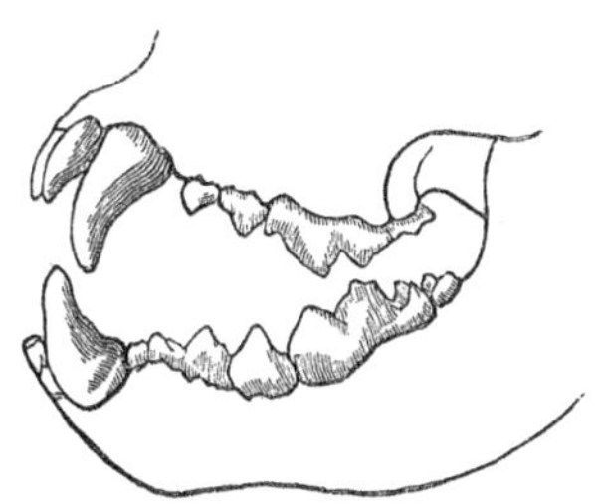

Fig. 42. — Teeth of a Flesh-Eater.

CATS.

Of all the Carnivores the Cats have the keenest senses, the quickest movements, and they are the most rapacious. Their tread is noiseless, — the bottoms of their feet being like a cushion; they stealthily approach their prey, and when near enough seize it with a sudden spring. The name *Cat* is not only given to

Fig. 43. — Puma.

the domestic varieties of this sort, but also to the Lion, Tiger, Panther, Leopard, Puma, Lynx, Jaguar, and Wild-Cat. The Lion, Panther, and Leopard inhabit Africa and Southern Asia, and the Tiger is found in India; the first and last being the largest of all the Cat tribe. The Puma is found from Canada to Patagonia, and is larger than the largest dog, and preys upon deer, sheep, hares, and sometimes attacks human beings. It climbs trees, and often lies upon a limb in wait for prey. The Jaguar inhabits Texas, and is found as far south as Patagonia. The American Wild-Cat and Canada Lynx much resemble each other, but the Lynx is the larger, being about three feet long, and the

Fig. 44. — Canada Lynx.

ears are tipped with long black hairs. They feed upon small quadrupeds and birds, sometimes pursuing the latter into the tops of trees.

HYENAS.

Hyenas live in Africa and Asia, and are about the size of a very large dog. They live in dens and caves, coming forth at night in search of food, feeding mainly on animals which they find dead. They are ferocious and greedy, and have such stout teeth and powerful jaws that they are able to crush the bones of the largest prey; and they swallow the fragments without masticating them.

DOGS, WOLVES, AND FOXES.

The Dog is the only animal that has followed man to all parts of the world. The varieties are numerous, and differ from one another greatly in their appearance and habits. Some of the most distinct varieties are the Greyhound, St. Bernard, Newfoundland, Esquimaux,

Shepherd's Dog, Fox-hound, Stag-hound, and Blood-hound, Spaniel, Setter, Pointer, Poodle, Terrier, Mastiff, &c. The Dog is noted for his sagacity, courage, and faithfulness, and if there were room many interesting stories might be related illustrating these qualities. Mrs. Lee tells the following story of a Pointer belonging to her father: "Clio stood with her hind legs upon a gate for more than two hours, with a nest of partridges close to her nose. She must have seen them as she jumped over the gate, and had she moved an inch they would have been frightened away. My father went on, and having other dogs did not miss Clio for a long time; at length he perceived she was not with the rest, and neither came to his call nor his whistle; he went back to seek her, and there she stood, just as she had got over the gate. His coming up disturbed the birds and he shot some of them; but Clio, when thus relieved, was so stiff that she could not move."

Wolves are ferocious and greedy animals, about the

Fig. 45. — American White and Gray Wolf.

size of a large dog. They often hunt in companies or packs, and thus are able to kill animals which singly they could not master. In newly settled parts of the country they destroy sheep, calves, and other animals of the farm. The White and Gray Wolf is found in nearly all the thinly settled regions of North America. The Prairie Wolf is common in the regions west of the Mississippi River.

Foxes are distinguished from all the rest of the Dog family by their pointed muzzle and large bushy tail. They are the most sly and crafty of all animals, contriving to steal turkeys, geese, chickens, and whatever they want to eat, and carry them away to their lurking-places in the woods and thickets. They are hunted with hounds which go in swift pursuit, while the hunter, knowing the habits of the animal, conceals himself in some valley or other locality where the fox will be almost sure to pass, and when he comes near enough shoots him down. But it must be stated that in many cases the shrewd movements of the fox deceive both the hunter and the dogs. If captured alive, which rarely happens, and struck while it is in a situation from which it cannot escape, the fox feigns itself dead, though unhurt, and when its captor is off his guard, will jump up and run away.

CIVETS.

Civets are about the size of the house cat, and with one exception belong to the Old World. The Civet-Cat of Texas and California is of a grayish color, its tail white with black rings. It lives upon the trees, is lively and playful, and, though shy, is easily tamed, and the miners often keep it as a pet.

Fig. 46. — Civet-Cat.

FISHERS, MARTENS OR SABLE, WEASELS, OTTERS, &c.

These animals have, in most cases, a slender body, long soft fur, especially in winter, and they are very quick and graceful in their movements, and exceedingly destructive to other small animals.

The American Fisher is about the size of a cat, but with a much more slender body, and is nearly black. The American Sable, or Pine Marten, of the Northern States and Canada, is much smaller than the Fisher, of a blackish brown color, and is celebrated for its beautiful and valuable fur, which is generally called the Hudson's Bay Sable. The fur known as the Russian Sable comes from a very similar animal which lives in Siberia. The Pine Marten delights in dense woods, where it pursues and captures hares, birds, and squirrels, swiftly following the latter even among the tree-tops. Its retreats, especially in winter, are hollow trees, and it is often seen by the hunter sitting with the head just out of its hole. If shot while in this position, it falls back into the hole and is lost; so the hunter, knowing its habits, walks slowly around the tree; the sable comes out to gratify its curiosity by a look at the hunter, and is then shot and falls to the ground. More than a hundred thousand skins of this animal have been collected in Northern North America in a single year.

Fig. 47. — Weasel.

Fig. 48. — American Sable.

True Weasels vary from five inches to a foot in length,

and are generally brown in summer and white in winter, the tail tipped with black. There are a half-dozen kinds in North America. The fur known as Ermine is furnished by the Weasels, the most valuable coming from Siberia. Weasels are generally bold, courageous, and extremely bloodthirsty, eagerly attacking animals much larger than themselves. They destroy rats and birds, and commit great havoc among poultry, a single individual having been known to kill fifty chickens in one night and the evening of the following day; and to kill several chickens in a coop near which a man was standing!

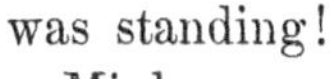

Fig. 49. — Mink.

Minks are about a foot and a half long to the tail, and are dark brown or black. They are found about ponds and streams, and their fur is very beautiful, and is often sold under the name of American Sable.

The Wolverine, found in the Northern States and

Fig. 50. — Wolverine.

Canada, and in the northern parts of Europe and Asia, is about three feet long, of a dark color, and is very powerful and ferocious when attacked. It is very troublesome to Sable hunters, breaking down their wooden traps, and eating the bait and game. It is so shrewd that it scarcely ever enters the trap, and hence one is seldom caught.

Otters live in and about the water, and feed upon fish. They are sportive in their disposition, and amuse themselves by "sliding down hill." Selecting a steep bank of a river, they slide head foremost into the water, and repeat the operation many times, apparently with great delight. Otters are three or four feet long from the nose to the tip of the tail, the color dark brown, and the fur is of two kinds, one short, fine, and thick, the other long, coarse, and scattered. When taken young, Otters are easily tamed, and become so familiar that they will lie in the lap like a cat.

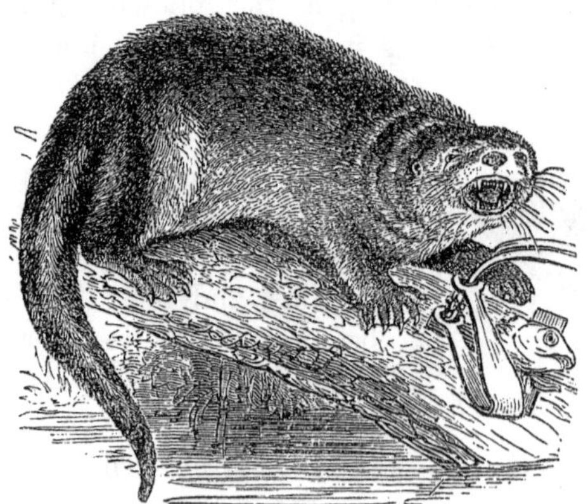

Fig. 51. — American Otter.

Fig. 52. — Skunk.

Skunks are found only in America, and are notorious on account of their disagreeable odor. They are a foot and a half long to the tail, and the color is black and white. They live in burrows, and seek their food at night, eating beetles and other small insects, and eggs.

The Badger of Western North America is about two feet long, with a stout body and

Fig. 53. — American Badger.

short tail, and its color is gray. The hair is long, extending on the hind part of the body so as nearly to conceal the tail. Badgers live in burrows, and dig with astonishing rapidity.

BEARS AND RACCOONS.

Bears and others of this family walk on the whole sole of the foot. They feed upon flesh, berries, and roots.

The Raccoon of the United States is about as large as a middle-sized dog, with a thick body, looking some-

Fig. 54. — Grizzly Bear.

Fig. 55. — Raccoon.

what like a little bear with a long tail; the color grayish, and the tail ringed with black and dingy white.

Bears are very large. The Grizzly, of the Rocky Mountains, is six or eight feet in length, and weighs in some cases eighteen hundred pounds, and the nails or claws are *six inches* long! It is the most powerful animal in America, and when wounded is very dangerous to the hunter. It has been seen to drag away a large bison, after killing it. The Black Bear of the Northern States is much smaller than the Grizzly, and less ferocious, seldom attacking men when not molested; but if disturbed when accompanied by its young, which are called cubs, it fights very savagely.

SEALS AND THE WALRUS.

The Seals and the Walrus live in the sea, but often come upon the rocks and ice-banks to lie in the sunshine. The head of the Seal much resembles that of

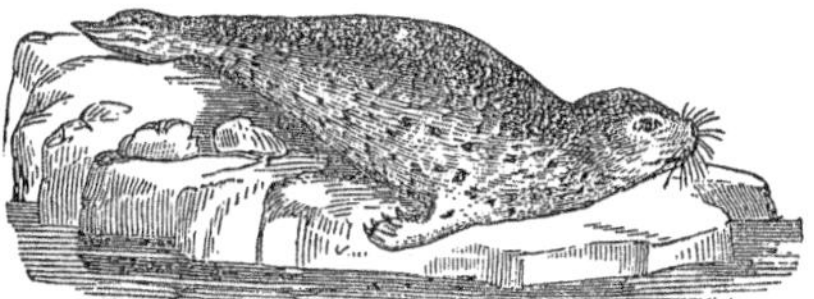
Fig. 56. — Seal.

a dog, and its eyes are beautiful and intelligent in appearance. When taken young, seals are easily tamed, and become attached and obedient to those who feed them, coming at call, and performing curious feats according to their master's directions. A few years ago, in a large tank of sea-water in the Aquarial Gardens at Boston were two Seals called "Ned" and "Fanny," which were so tame that they would come to the keeper at call, and allow him to handle them, would shoulder a miniature musket, turn the crank of a hand-organ, shake hands with the by-standers, and "Ned," especially, would even "throw a kiss" to the ladies. Seals feed upon fish, and always eat in the water. They are from three to twenty feet long. The Walrus has a body as large as the largest ox, and is covered with short brown hair. Two of its upper teeth, the canines, or eye-teeth, grow to be tusks two feet long. These tusks assist in climbing upon the ice-banks, and serve as a means of defence, and to aid in securing food.

HERBIVORES, OR PLANT-EATERS, OR HOOFED ANIMALS.

These are mammals which feed wholly upon vegetation, and which have hoofed feet, and use their limbs only for standing, walking, and running. Some of them, as the Deer, Antelopes, Sheep and Goats, and Oxen, have the foot divided or cleft, forming an even number of toes; and all herbivores of this sort chew the cud, and from the latter fact are known as Ruminants, a name which means *cud-chewers.* Others, as the Horse and Ass, have only one toe to each foot, and are hence called Solipedes or Solid-hoofed animals. Others, as Elephants and the Mastodon, have five toes, and a long flexible snout or proboscis, and from the latter fact are called Proboscidians.

DEER.

The Moose, Reindeer, Deer, and Elk all belong to the Deer family. The males have solid horns called antlers, which they shed once a year; new and larger ones growing to take the places of those which have been shed.

The Moose is the largest of all the Deer kind, being as large as a horse, and with an exceedingly long head,

Fig. 57. — Moose.

large flattened horns, and very long legs. It travels with an awkward gait, but with great speed, easily making its way through deep snows, bushes, over brush-heaps, fallen trees, fences, and whatever obstructions lie in its path. It is quite common in some parts of Maine, Northern New York, and Canada. Color grayish brown.

The Reindeer is a much smaller animal than the Moose, being about five feet long and three feet high. It has become celebrated for the services it renders the Laplanders, who keep large herds of Reindeer, using them for beasts of burden and for drawing their sledges, — a sort of sled, — their milk and flesh for food, and their skins for clothing. They are very hardy animals,

Fig. 58. — American Reindeer, or Caribou.

and subsist on the coarsest fare, eating the tender portions of shrubs in summer, and in winter scraping the snow from the ground and feeding upon the "Reindeer-moss." The American Reindeer, or Caribou, of Maine and Canada, and other northern parts of North America, is by some thought to be of the same kind as the one found in Lapland. Unlike their relatives, both the male and female Reindeer have horns.

The American Elk, or Wapiti, is another kind of Deer which lives in the wooded regions of the northern parts of North America, and which is about as large as the Moose, and has horns five or six feet long, and very much branched.

Fig. 59. — American Elk, or Wapiti.

The Common Deer, of the wild regions of the United States, is one of the most beautiful and graceful of all its family. It is very timid, and, when alarmed, bounds swiftly away. It is about the size of a sheep, but with a much more slender body and much longer legs. It is hunted in the autumn and winter, and great numbers are sent to the markets. Its flesh is called venison, and is highly prized for food.

Fig. 60. — Common or Virginia Deer.

The Musk Deer inhabits Thibet, and is smaller than the Common Deer, and has no horns. In each side of the upper jaw are long canine or eye teeth, like tusks. The musk used in making perfumery is furnished by this animal. It is contained in a pouch, or sack, on the under side of the body.

Fig. 61.— Musk Deer.

ANTELOPES.

Antelopes are found in Europe, Asia, Africa, and North America, but are most numerous in Southern

Africa, where there are many kinds, and where herds of ten thousand or more are sometimes seen together. Their horns are round, variously wrinkled and curved, and black. Antelopes vary in size from those as small as a deer to those as large as a horse. The large kinds belong to Africa.

The Pronghorn Antelope, of the Rocky Mountains, is larger than a sheep, with much longer neck and legs. Its hair is coarse and thick. It gets its name from the

Fig. 62.—Pronghorn Antelope.

prong, or branch, on each horn. This animal is found at times in large numbers, herds of a thousand and more having often been seen.

The Mountain Goat, of the Rocky Mountains, is an antelope, and not a true goat, as one would suppose from its name. It is entirely white, except its horns and hoofs, which are black. Its fleece is long and very fine, being equal in quality to that of the celebrated

Cashmere Goat. It inhabits the lofty peaks of the mountains, frequenting the steepest places.

Fig. 63. — Rocky Mountain Goat.

The Gazelle, of Africa and Asia, is about the size of a small deer, and is celebrated for its beautiful and grace-

Fig. 64. — Gazelle.

Fig. 65. — Chamois.

ful form, and for its large, dark, and lustrous eyes. The Orientals, or inhabitants of the East, compliment a lady by comparing her eyes to those of the Gazelle. When taken young, though wild and timid, it is easily tamed, and becomes a great favorite. The Chamois, of the high mountains of Western Europe, is about the size of a goat, of a dark brown color, and its horns, towards the summit, are bent backwards like a hook. It is very shy, and on the slightest alarm bounds swiftly away over rocks, glaciers, along dizzy heights, up and down precipices, where it would seem no animal could get a foothold, often leaping upon a rock just large enough to receive its four feet placed close together.

SHEEP AND GOATS.

Sheep have the horns angular and directed back-

Fig. 66. — Mountain Sheep, or Big-Horn.

ward, then spirally curved forward, and yellowish-brown in color, instead of round and black, as in Antelopes. The Mountain Sheep, or Big-Horn, of the Rocky Mountains, is much larger than the domestic sheep, and with very large horns. The hair is of a gray color and very coarse. The hunters say that this animal will leap from a height of fifty feet and strike upon the tips of the spiral horns, receiving no injury.

Goats have the horns directed upward and backward, and the chin usually has a long beard. The wild kinds live upon the high and rugged mountains of Asia. The Wild Goat of Persia is supposed to be the parent of the common domestic goat. The Cashmere Goat of Thibet is celebrated for its fine wool. Its hair is long and silky, and under this is a delicate gray wool, of which the costly Cashmere shawls are made. Three ounces are obtained from a single animal.

OXEN.

The Musk Ox, of Arctic America, is of the size of a small cow, with very long dark-brown silky hair. It

Fig. 67. — Musk Ox.

feeds upon grass in the mild season, and in winter upon

mosses and lichens, which it gets from the steep sides of hills that are blown bare by the winds, and up which it climbs with the agility of the chamois.

The Bison, or Buffalo, of the Western plains, is the largest quadruped of America, being of the size of a large ox. It is covered with a thick coat of dark hair, that about the head and shoulders being long and

Fig. 68. — Bison, or American Buffalo.

shaggy. At the time of the discovery of America the Buffalo was found even to the shores of the Atlantic, but it has been driven back until it is now found only beyond the Missouri and the head-waters of the Mississippi. Here it is not uncommon to see the prairies covered with Buffaloes as far as the eye can reach; and travellers have passed through herds of them for days and days in succession, with scarcely any apparent lessening of their numbers. Their paths resemble travelled roads; and as their routes, in most cases, extend in a straight line from one convenient crossing-place of a river or ravine to another, taking springs or streams in their course, they frequently serve as the highways of travel across the prairies. Though naturally timid, the Buffalo, when wounded, is furious, and dangerous to the hunter. It is estimated that five hundred thousand of these animals are killed every year; many being slaughtered merely for sport, or perhaps for the sake of the tongue alone, but most of them for their skins, which make the well-known buffalo-robes.

CAMELS AND LLAMAS.

The Camel is a native of Central and Southern Asia, and, from the earliest times, has rendered such impor-

Fig. 69.—Llama.

tant services to the inhabitants of the East in carrying merchandise across the deserts, that it has been called the "Ship of the Desert." Its feet are fitted for travelling in the sand, its strength and power of endurance are very great, and it can live on the coarsest and most scanty vegetation, and travel for days without drinking. It can carry from five hundred to one thousand pounds, and kneels to receive and to be relieved of its load.

The Camel is larger than the horse, and stands very high. There are two kinds,—one with two large humps upon the back, and the other with only one hump.

The Llamas inhabit the Andes of South America, and are much smaller than the Camel, being only four or five feet high, and they have no hump. They live in herds, and are tamed and used as beasts of burden. The Alpaca is a variety of Llama with long woolly hair, which furnishes material for valuable fabrics.

WHALES, OR CETACEANS.

These mammals live in the water, have their limbs paddle-like and fitted for swimming, and their whole appearance is fish-like; but they are true mammals, nourishing their young with milk, breathing air for which they come often to the surface of the water, and their blood is warm. Most of them are large, and some of them are the largest of living animals, and they are covered with a smooth skin. They breathe through a hole, or holes, on the top of the back part of the head, and through these some kinds blow or spout water to the height of thirty, and sometimes even to fifty feet.

RIGHT AND SPERM WHALES.

The Greenland or Right Whale attains the length of sixty or seventy feet. It has no real teeth, but in the upper jaw are rows of upright horny plates, called whalebone, which are fringed on their inner edges. Its food is small marine animals. Swimming through schools of these, the Whale takes millions into

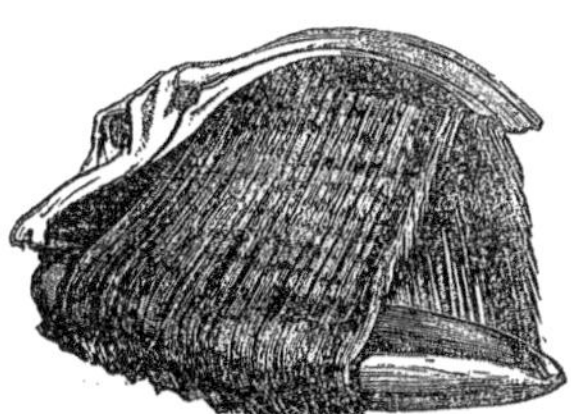

Fig. 70.—Skull of the Right Whale, showing the whalebone.

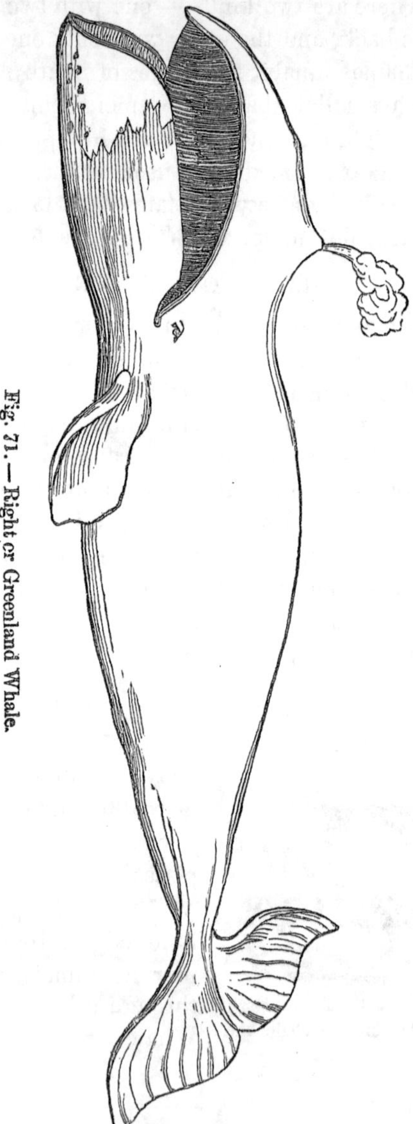

Fig. 71.—Right or Greenland Whale.

his mouth at once. This Whale supplies the world with whalebone, and also furnishes more oil than any other. Its home is in cool and frigid seas.

The Great Sperm Whale, of the warm parts of the ocean, is fully equal to the Right Whale in size. The upper jaw has neither teeth nor whalebone, but the

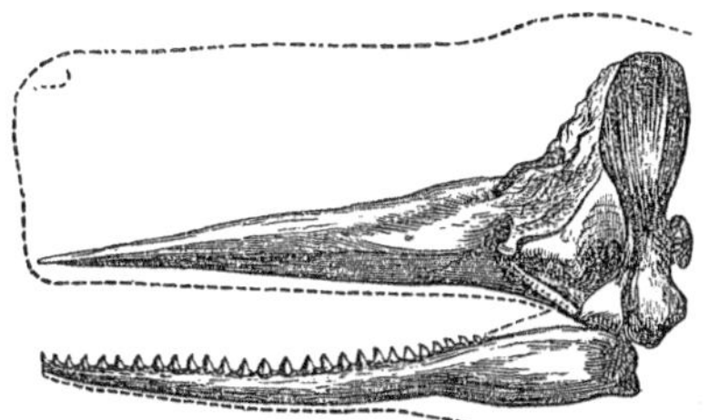

Fig. 72. — Head of Sperm Whale.

lower has teeth. In the upper portion of the head there are cavities filled with oil which hardens when cool, and is known as spermaceti. The body yields sperm oil. Ambergris, a substance used by chemists in making perfumery, is found in the intestines of this whale.

The spouting, or blowing, is different in these two whales; for the Right Whale has two blow-holes on the top of the head, and spouts water as well as the warm moist air of the lungs; while the Sperm Whale has only one blow-hole, and spouts only the moist breath of the lungs, which, on contact with the air, forms a white mist that instantly vanishes.

Through the kindness of Hon. William Mitchell, Mr. Tenney has been permitted to make very interesting extracts from letters written by Captain William Barney of Nantucket, illustrating the habits of Whales. One of these extracts is here given, which shows how quickly the Sperm Whales find out when their companions are in difficulty.

"I have been looking over in my mind some of my voyages in the Pacific, and one circumstance is now fresh in mind. In the year 1824, when I was second mate of the ship Maria, of New Bedford, while cruising near the Marquesas, or Washington Islands, the lookout at the masthead saw a large Sperm Whale off our lee bow, about three fourths of a mile distant, going with moderate speed the same course the ship was then steering. There was no other whale then in sight. The ship was laid aback, the boats were lowered, and the chief mate and myself went in pursuit. The whale went down before we reached him; and while he was down a signal was made from the ship that another whale was in sight, two miles ahead of the ship, and going the same course. Soon the whale which we were in pursuit of came up, and the mate pulled on and struck him. I pulled up to assist the mate, when a signal was made from the ship that the whale ahead was coming towards the boats and the struck whale. I left the fast whale with the mate, and prepared to receive the other. Soon we saw him coming at the top of his speed, or 'eyes out,' as the whalemen say. He came directly to the fast whale, lashing the water with his flukes, and floundering around violently. I soon got an opportunity, and struck him, whereupon he turned upon my boat and tried to stave it to pieces, but after a hard struggle I succeeded in conquering him. When we got to the ship we found that the instant the first whale was struck by the mate, the other whale ahead of the ship struck his flukes in the water and disappeared, but in a few moments was up again, and coming with great speed towards the fast whale and the boats. No doubt that when the first whale was struck he perceived it, and turned from his course and came to assist or defend his partner.

"Two schools of whales may be two or three miles apart, and both thrashing the water with their tails, and apparently taking no notice of each other; but let one of either school be struck, and those of the other seem to know it instantly."

DOLPHINS, PORPOISES, AND WHITE WHALE.

These animals live in herds, and prey upon fishes. The Common Dolphin is about eight feet long, black

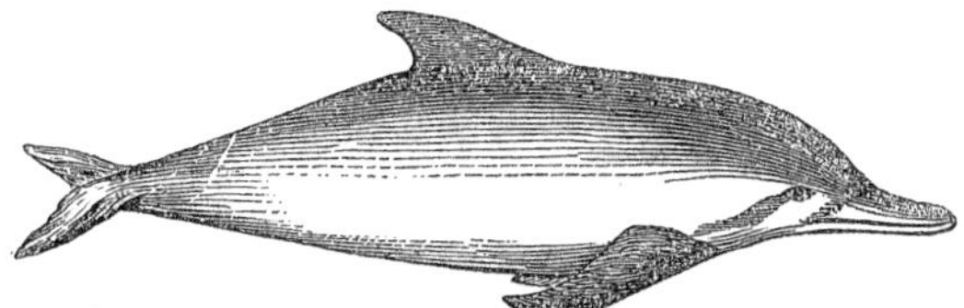

Fig. 73.—Dolphin.

above and white below. The ancients believed this animal to be very docile and fond of music. The White Whale lives in the Northern Seas, and is from

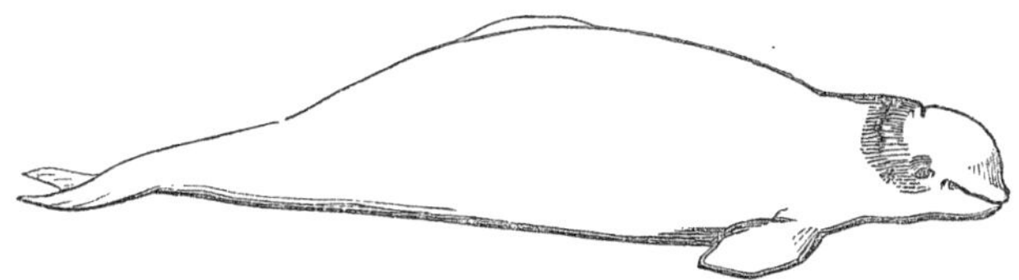

Fig. 74.—White Whale.

ten to twenty feet long. It often ascends rivers, and is frequently seen in the St. Lawrence. One of these animals, about ten feet long, was kept for two years in the Aquarial Gardens in Boston. He was quite docile, knew his keeper, and would come and take food from his hand. He was trained to a harness, and drew a young lady in a car prepared for the purpose.

The mammals already described are mostly of large size; we now come to the smaller ones.

BATS, OR CHEIROPTERS.

Bats are animals which have a thin skin reaching from the neck to the hind legs, and extending to the ends of their long fingers. By means of this skin they can fly as easily as birds, and their flight is noiseless

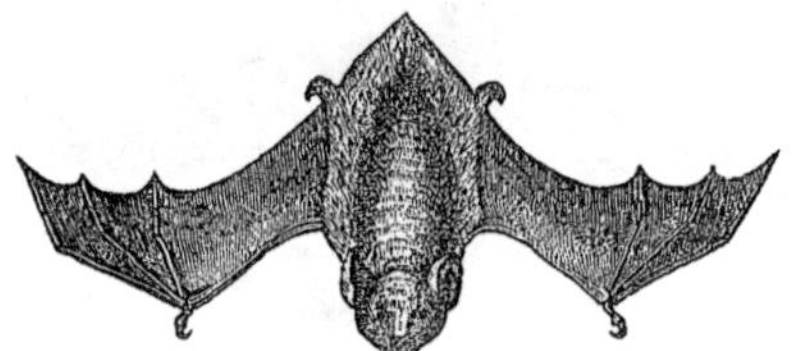

Fig. 75. — Hoary Bat.

and rapid. The body is covered with soft fur. Their eyes are very small, ears large, and the thumb has a sharp hook. In the daytime they stay in caves, hollow trees, or other dark places, hanging by their hooks, or by the sharp claws of their hind feet. Bats can fly through the most winding and crooked passages without harm, even after their eyes have been destroyed. Some of the larger ones of the East Indies eat fruits and birds, but most kinds feed upon insects, which they are catching when we see them flitting and turning hither and thither in the dusk of evening. The Red, and the Hoary Bat, three or four inches long, are common species in North America.

INSECT-EATERS, OR INSECTIVORES.

The Insect-Eaters include the Shrews, Moles, and Hedgehogs. Many naturalists also place here the Gale-

Fig. 76. — Galeopithecus.

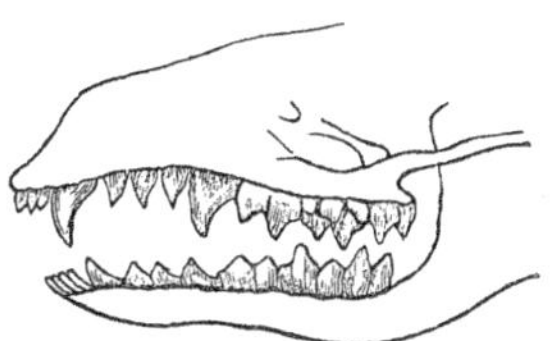

Fig. 77. — Teeth of an Insect-Eater.

opithecus, a curious bat-like animal found on trees in the Indian Archipelago. Insectivores sleep during the day, and go forth at night in search of food. In cool regions, many of them sleep all winter.

SHREWS.

Shrews are little mouse-like animals, — but smaller than the smallest mice, — with a long and tapering head and soft silky fur. They live under rubbish, and

Fig. 78. — Thompson's Shrew.

Fig. 79. — Water Shrew.

in holes which they dig in the ground. They are very quarrelsome; and if two are confined together the weaker is soon killed. There are more than a dozen kinds in North America.

MOLES.

Moles have a stout, thick body, short, strong legs, short tail, and very large fore feet fitted for digging. Their eyes are very small, and their fur is soft, thick, and velvet-like. The Shrew Mole of North America is of the size of a very large mouse, and its eyes are so small that many suppose it to be blind. The hole for the eye is only about the size of a hair, and the eyeballs are smaller than a mustard seed. It is said that this mole comes to the surface of the ground every day at noon. The Star-nosed Mole is about the size of the Shrew Mole, and is so named from the form of the end of the nose, which is star-shaped.

Fig. 80.—Shrew Mole.

Fig. 81.—Nose of Star-nosed Mole.

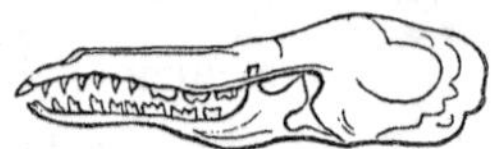

Fig. 82.—Skull of Star-nosed Mole.

HEDGEHOGS.

These animals are short and thick, and the back is covered with spines. When alarmed, they take the form of a ball, presenting the spines in every direction,

Fig. 83.—Madagascar Hedgehog, or Tenrec.

to ward off attacks. They sleep during the day in concealed places, and come forth at night to feed upon in-

Fig. 84.—European Hedgehog.

sects, fruits, and roots. In cold climates they sleep all winter. They live in the Old World. They are all small, the European Hedgehog being only nine or ten inches long. The animal in America which is called Hedgehog is a Porcupine.

RODENTS, OR GNAWERS.

The Rodents are readily known by their teeth. In each jaw they have the two front ones chisel-shaped, and between these and the grinders there is a wide space without teeth. The front teeth wear in such a manner that the more they are used the sharper they become, and they grow at the base as fast as they wear away at the top. More than six hundred kinds of Rodents are known, most of which are small; the Beaver, with one exception, being the largest.

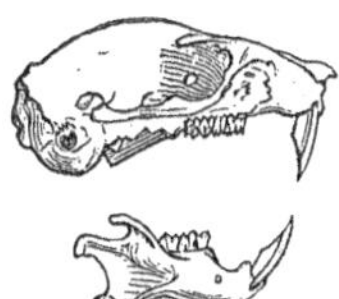

Fig. 85.—Skull of a Rodent.

The Rodents include the Squirrels, Gophers, Woodchucks, Rats and Mice, Porcupines, Hares, &c.

SQUIRRELS.

Squirrels are small and very pretty animals, with large bright eyes, long ears, divided upper lip, and long bushy tail. They are lightly built, agile, and live upon trees, and feed upon fruits and nuts. There are about fifty kinds in America, and twelve or more in the United States. The most prominent kinds are the large Fox Squirrels of the Middle, Southern, and Western States, and the well-known Gray, the Red, and the Flying Squirrels found over a large part of the United States. Gray Squirrels are noted for their occasional extensive migrations. Assembling in large numbers, they make their way across the country, swimming rivers, and turning aside for no obstacle. Gray squirrels occur of every shade from gray to jet black.

Fig. 86. — Gray Squirrel.

The Red Squirrel is seen at all seasons and in all weathers. In the Northern forests the deepest snows of winter are soon covered with its tracks, and penetrated by holes bored to find the cones of spruce and pine, and the nuts scattered beneath, or hidden the previous autumn. It often sits for hours upon a stump or limb of a tree close to the trunk, and, holding a cone or nut in its fore paws, gnaws it briskly till it gets all the food it contains. If disturbed while upon the ground, this squirrel runs up the nearest tree, leaping from branch to branch, and from these to another tree, and soon passes out of sight. Sometimes, when startled, it commences chattering with great fury, and leaping about as if in defiance of the intruder.

The Flying Squirrels have a thin skin, or membrane, covered with fur, which extends along the sides of the body between the fore and hind legs, and which, when spread out, serves as a support in leaping from tree to tree, and enables them to perform a sort of flight. They are nocturnal, and therefore not often seen. Their nests are made in the hollows of trees, where large companies often live together. The Common Flying Squirrel of the United States is about five inches long, and the fur is soft, silky, and yellowish brown. It is quite easily tamed, and, being gentle and very beautiful, makes a pleasant pet.

Fig. 87. — Flying Squirrel.

The Striped Squirrels have cheek-pouches, in which they carry grain and nuts to their holes, and they have a shorter and less bushy tail than the others. The

Fig. 88. — Striped Squirrel, or Chipmunk.

Common Striped Squirrel, or Chipmunk, is about five inches long to the tail, and the color is yellowish gray, with five black stripes on the back and sides. In autumn the Chipmunks may be seen with their cheek-pouches full of nuts or grain, which they store up for their food in winter, at which time they always remain in their holes.

The Striped Gopher, of Michigan and southward, is a very beautiful animal, about the size of the Red Squirrel, of a dark brown color, with light lines and rows of light spots. It lives in burrows, and when alarmed pops into its hole with a chirp. The Prairie Dog is larger

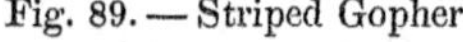
Fig. 89. — Striped Gopher.

Fig. 90. — Prairie Dog.

than the Striped Gopher, appearing somewhat like a small woodchuck. It utters a sharp chirp, called barking, and hence its name. It lives in burrows, and large numbers are found in the same locality, forming communities called "Dogtowns." Before each hole is a little hill of earth, upon which is almost always a Prairie Dog on the lookout for intruders, and upon the slightest alarm it dives into its hole, but soon appears again. Their holes are also the home of the Burrowing Owls and of Rattlesnakes.

Beavers are about three feet long to the tail, and are the largest of the Rodents, excepting an animal called the Capybara which lives about the rivers of South

America. Beavers have a flat, scaly tail, and are wholly aquatic in their habits, and their food is chiefly bark and aquatic plants. Their teeth are very sharp and

Fig. 91.—American Beaver.

powerful, enabling them to gnaw down trees of the hardest wood. Beavers prefer running water, in order that the wood which they cut may be carried to the spot where it is to be used. They keep the water at a given height by dams, which they build of trees and branches mixed with stones and mud; and they build winter houses with the same materials. Each house consists of two stories; the upper story is above water and dry, and serves as a shelter; the lower is beneath the water, and contains their stores of bark and roots. The only opening to the hut is beneath the surface of the water. The color of the beaver is reddish-brown, and the fur is soft and fine. It lives in the unsettled parts of North America.

The Pocket Gopher, Pouched Rat, or Geomys, of the prairies of the Western States, is nine or ten inches long, with large front teeth, strong fore feet, and short tail. Opening on the outside of the mouth are large

Fig. 92.—Pocket Gopher.

cheek-pouches, which reach back even to the shoulders; and these pouches are lined with fur, and are entirely different from the much smaller cheek-pouches of the Striped Gopher, which open within the mouth. The Pocket Gopher throws up a mound of earth which, in some instances, is ten feet in diameter, and two feet high; and within this mound is its nest, where it rears its young; and from the mound it digs numerous galleries in different directions, one or two feet below the surface of the ground. It uses its curious pouches for carrying food, and for carrying away the earth which it removes in digging its galleries. Coming to the surface with its pouches full of earth, it empties them so quickly as to puzzle the looker on, and instantly retreats into its hole. Pocket Gophers feed mainly upon the roots of plants. They fight savagely with one another, and offer battle when met by man. If two are placed together, they instantly attack each other, and the stronger eats up the weaker.

RATS AND MICE.

There are more than three hundred kinds of these animals, all of which are small. More than fifty kinds inhabit North America. They devour all sorts of edible substances, animal as well as vegetable, and some even attack living animals.

The largest, except the Muskrat, is the Norway, Brown, or Wharf Rat, originally from Asia, but now exceedingly abundant in Europe and in this country. The Black Rat, which was introduced into this country from Europe more than three hundred years ago, is nearly as large as the Brown, and was formerly the most common large rat in stores, houses, barns, and other buildings, but is now rapidly disappearing before its more powerful rival, the Brown Rat, which pursues it, captures it, and even devours it. If the two kinds be placed together in a cage, the brown rats are sure to feast upon their darker companions. If one of their own number gets wounded, instead of aiding him, they fall upon and devour him. The Roof Rat, of the Southern States, originally from Egypt, where it lives in the thatched roofs of the houses, the House Mouse, originally from Asia, but now found in all countries,

Fig. 93. — White-footed Mouse.

the Harvest Mice, the White-footed Mice, the Field Mice, and the Jumping Mice, are other kinds which are found in the United States, but which cannot be described here for want of room. For further description, see Tenney's Manual of Zoölogy. The Jumping Mouse, however, is too interesting to be omitted. It is found over a large part of North America, and is about three inches long to the tail, which in some instances is even six inches in length; and the color yellowish-brown, lined with black, the lower parts white. It moves

Fig. 94. — American Jumping Mouse.

by very long and rapid leaps. It is found in the meadows and grain-fields.

The Muskrat, mentioned above, is very common about ponds, rivers, and brooks in North America. It is a foot long, besides the tail, which is about as long as the body, and the color is dark brown above and rusty brown below. The fur is now sold under the name of River Sable, and is much used for collars and muffs. Muskrats build winter houses of mud, sticks, and grass, the entrance being beneath the water, and leading to a dry apartment above.

PORCUPINES.

Porcupines are distinguished from all other Rodents by their spines, or quills, which are very sharp. The North American Porcupine is about two feet long, the color brown with long white-tipped hairs, and the tail and upper parts are covered with white spines. It lives in hollow trees and in holes among the rocks, and readily climbs trees. It eats bark, leaves, and green corn. It is often called the Hedgehog. See Figure 95. The Crested Porcupine, of Southern Europe, has quills nearly a foot long. These quills are used for pen holders.

Fig. 95.—American Porcupine.

HARES.

Hares are found in nearly all countries. In America there are about twenty kinds. They are timid, and have a habit of stamping with the hind feet when alarmed. The Common Hare, or White Rabbit, about twenty inches long, is brown in summer, and white in winter. It lives in the thick swamps, rarely enters holes when pursued, but depends for safety upon its fleetness. It follows the same paths year after year, both in winter and summer. The Gray Rabbit is a smaller kind, which does not turn white in winter.

EDENTATES.

The Edentates are Sloths, Armadillos, and Ant-eaters. Some of these animals have no teeth, and others are only destitute of front teeth. Many of them have a bony or scaly covering. They live in warm countries.

ARMADILLOS.

The word Armadillo means *clad in armor*, and is given to these animals on account of their bony or horny covering. They live in the warm and hot parts

Fig. 96.—Nine-banded Armadillo.

of America, dig burrows, and feed upon vegetables, insects, and worms. The Nine-banded Armadillo is about two feet long, and is found as far north as Texas.

MARSUPIALS.

The Marsupials have a pouch, or sack, beneath the body, in which the young are kept for a time after they are born, and even after they are able to walk they resort to the pouch of the mother when danger is near. With the exception of the Opossums of America, all the Marsupials are found in Australia.

OPOSSUMS.

Opossums are small animals, the largest being scarcely larger than the common cat, and the smallest but little larger than a mouse. They feed upon birds, bird's eggs, insects, and other small animals. The tail is long and is capable of being twisted around objects, thus aiding in climbing. The Opossum of the United States is about the size of a cat, the hair whitish with brown tips.

Fig. 97.—Opossum.

It often lies motionless for hours in the warm sunshine. When slightly wounded it has the habit of feigning itself dead, or "playing possum," and in that way often escapes from the inexperienced hunter.

KANGAROOS.

Kangaroos are marsupials which are remarkable for the great development of their hinder parts, — the hind

Fig. 98. — Kangaroo.

legs and tail being very long and powerful, and the fore legs very short, weak, and but little used in locomotion, which is accomplished by leaps of enormous extent. They live in troops, feed upon vegetation, and are harmless and easily tamed. They vary in size from that of a rabbit to that of a deer.

Fig. 99. — Wombat.

Fig. 100. — Skull of Wombat.

The Wombat is a very curious animal of Australia,

about the size of a woodchuck, and which, in its structure, resembles both the Rodents and the Marsupials. Its body is thin, legs short, and the tail is wanting. The Wombat feeds upon grass, and burrows in the ground.

DUCKBILLS, OR MONOTREMES.

These are animals which vary much from all other mammals, having their organic structure in some respects much like that of Birds. They belong to Australia. One of the most interesting kinds is called the

Fig. 101.—Duckbill, or Platypus.

Duckbill, or sometimes Platypus. Its muzzle is flat and appears very much like that of a Duck, its legs short, feet webbed, and its body is covered with short brown fur. It is less than two feet long, lives about ponds and streams, and digs burrows in the banks.

BIRDS.

Of all animals perhaps none are more interesting to both young and old than Birds. Their presence in the fields and hedges, in the groves and forests, their beautiful, and, in many kinds, splendid colors, their sweet songs, and their curious and wonderful habits, charm and delight every one.

Birds are vertebrates which are covered with feathers, furnished with a bill, and fitted for flight, — their form as well as their structure being adapted for easy and rapid movement through the air; even their bones are

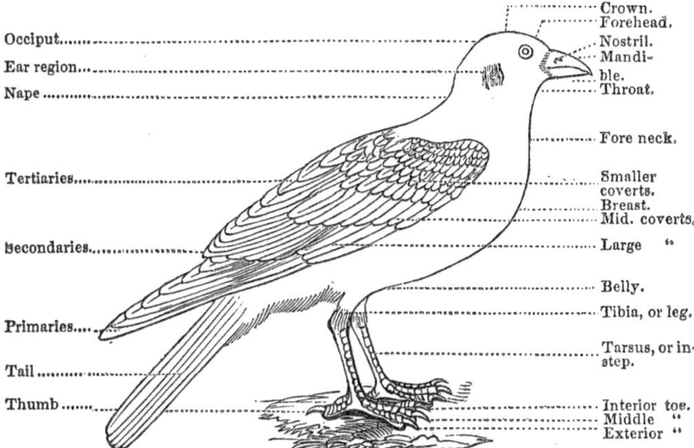

Fig. 102. — Showing the names of some of the principal parts of a Bird.

hollow, hence very light in proportion to their size. The general form of a Bird, and the names of some of the principal external parts, are shown in Figure 102. The skeleton and the names of the principal parts are shown in Figure 103. It is an interesting fact that the form and the skeleton of a Bird suggested the right way in which to build a ship in order to combine strength with swiftness.

Although the body of Birds is covered with feathers, these do not grow from the whole surface, but are arranged in rows and patches, with bare spaces between. Feathers are made up of a hard central portion or shaft, and a vane, the latter being the broad portion which con-

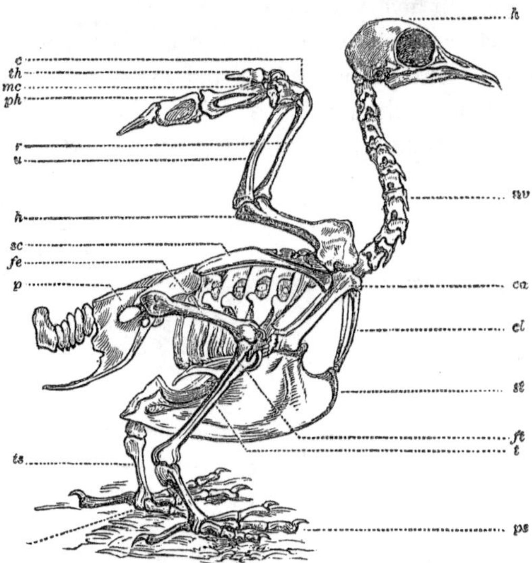

Fig. 103. — Skeleton of a Bird.

h, head; *nv*, neck vertebræ; *c*, wrist; *th*, thumb; *mc*, metacarpus, or hand; *ph*, phalanges, or fingers; *r*, radius; *u*, ulna; *h*, humerus; *sc*, scapula, or shoulder-blade; *cd*, corocoid bone; *cl*, clavicle, or "wish-bone"; *st*, breast bone, or sternum; *fe*, femur, or thigh bone; *p*, pelvis; *ft*, fibula and tibia united; *t*, tibia, or leg; *ts*, tarsus, or instep; *ps*, phalanges, or toes.

sists of delicate plates that are united by minute barbs along their edges, and thus made firm, — the plates not separating from one another when pressed against the air, as in flying. There are, however, downy feathers on every bird, or such as do not have the plates united. The plumage of Birds is made water-proof by the oil with which they dress their feathers, and which is obtained from a gland situated on the tail. They shed their feathers twice a year, and in many kinds the winter plumage differs in color from that of the summer. In most birds the colors of the male are much more brilliant than those of the female.

Birds swallow their food without chewing it, and it is first received into a sack called the crop; then it passes into another sack, where it is moistened and softened; then it passes to the gizzard, where it is digested. The gizzard generally contains gravel and other hard substances, which these animals swallow to aid digestion.

Birds lay eggs and sit upon them to hatch them, and most birds build nests in which to rear their young, those of the same kind building alike. The young bird in the egg has a horny point at the end of the bill, with which it breaks the shell. This point is plainly seen on the bill of the newly-hatched chicken; in a few days it falls off.

The number of kinds of Birds is ten or twelve thousand, and there are about seven hundred kinds in North America. Birds of Prey, the Climbers, the Perchers, the Scratchers, the Runners, the Waders, and the Swimmers are the large groups into which Birds are divided.

BIRDS OF PREY, OR RAPTORES.

These are the Vultures, Eagles, Hawks, Falcons, and Owls. Most of them capture birds and other animals for food. They are mostly of large size, and have a strong hooked bill, sharp claws, great spread of wing, and very powerful muscles, and the females are generally larger than the males. They live in pairs, and choose their mates for life.

VULTURES.

Vultures have the head nearly naked or thinly cov-

Fig. 104.— California Vulture.

ered with feathers, and, unlike the other rapacious birds, seldom capture prey, but feed upon dead and decaying animals, which they trace by sight or scent at great distances. They make no nest, but deposit their eggs on the ground or naked rock. There are three or four kinds in the United States. The celebrated Condor of the Andes, and the Lammergeyer of the Alps, are vultures of the largest kind. The latter attacks lambs, goats, and the chamois. The California Vulture is the largest bird of prey in North America, being as large as the largest Turkey; the color is black, the head orange and red. See Figure 104.

FALCONS, HAWKS, AND EAGLES.

These birds have the head clothed with feathers, and their talons are very sharp. Their flight is rapid, and they attack their prey with great ferocity, capturing chickens, ducks, grouse, quails, hares, rabbits, squirrels, and other small animals. The species are numerous, about seventy kinds of Eagles being known in all countries; and more than thirty kinds of Falcons and Hawks inhabit North America. The true Falcons have a distinct tooth in the upper mandible, as seen in Figure 105.

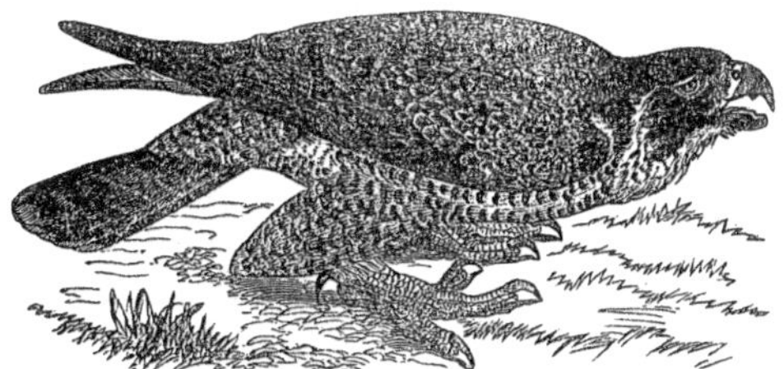

Fig. 105.—American Peregrine Falcon, or Duck Hawk.

The Duck Hawk, or Peregrine Falcon, of North America, pursues its prey with almost inconceivable

Fig. 106.— White-headed or Bald Eagle.

velocity through all its turnings and windings, and when within a few feet, protrudes its talons, grasps the prize, and bears it away to some secluded place and devours it. Sometimes it sweeps over the water and catches up ducks and other swimming birds. This falcon is about a foot and a half in length. The Peregrine Falcon of Europe, very much like this species, was formerly much used in falconry, a fashionable sport of kings, nobles, and fair ladies.

Fig. 107.— Sparrow Hawk.

The Sparrow Hawk, of America, is the smallest of the hawks, being but little larger than the common robin. It preys upon small birds, mice, and insects. It becomes attached to a particular locality, and may be seen day after day on the same tree or stump watching for prey.

The Bald, or White-headed Eagle, of North America, is found along the sea-coasts, lakes, and rivers, and usually makes its nest on some tall tree. Although called Bald, its head is clothed with white feathers. Its principal food is fish, which it obtains mainly by robbing the Osprey, or Fish-Hawk. Seated on a dead limb of a large tree that commands a view of the waters, it watches the Fish-Hawk as he descends and plunges into the deep, and, as he emerges with his prey and rises into the air, the Eagle gives chase; each moves with its utmost speed, but the Eagle rapidly gains, and as it is about to reach the Hawk, the latter drops the fish; the Eagle sweeps downward, snatches it before it reaches the water, and bears it away to the woods.

OWLS.

Owls are birds of prey which, in most cases, are active by night, and rest during the day. Their large head, and large staring eyes, and the tufts of feathers resembling ears, which many of them have, give to the face a strange, cat-like expression. Their plumage is soft and loose, and their flight is almost noiseless.

Fig. 108. — Great Horned Owl.

They prey upon birds, hares, squirrels, mice, and insects. There are about forty kinds of Owls in America, varying from the size of a robin to that of a small turkey. The Great Horned Owl has large ear-tufts standing up like horns; the Screech Owl is small, and is noted for its tremulous, doleful notes; the Long-eared Owl has very long ear-tufts, and its cry is prolonged and plaintive, consisting of two or three notes repeated at intervals; the Gray Owls are very large; the Saw Whet Owl is small, and its notes sound like the noise made in filing a saw; the Burrowing Owls are very small, and live in the burrows of the Prairie Dog; and the Snowy Owl is large, and, unlike the preceding ones, hunts in the daytime as well as at twilight. It lives in the cold regions, and is seen in the United States only in winter.

Fig. 109. — Snowy Owl.

CLIMBERS, OR SCANSORES.

These birds have the toes in pairs, two in front and two behind. Parrots, Cuckoos, and Woodpeckers are the principal kinds.

PARROTS.

Parrots have a stout, thick bill, hooked at the tip. Many of them are adorned with the most gorgeous-colored plumage; and this, together with the ease with which they are trained to speak, has made them objects of great interest. They live in the warm regions, where the trees are always green, and fruits and seeds never fail them.

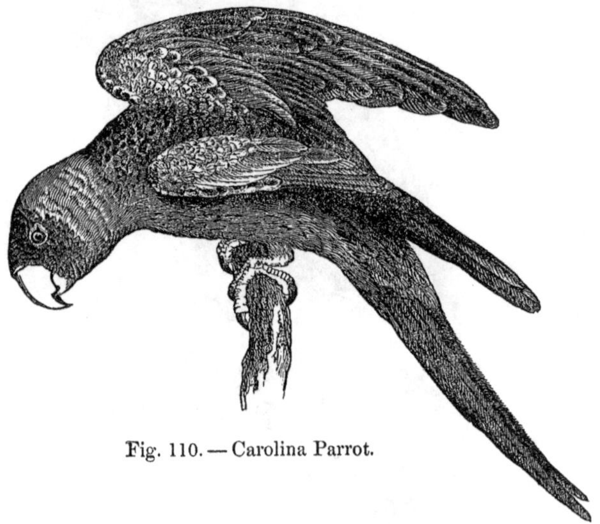

Fig. 110. — Carolina Parrot.

The Carolina Parrot of the Southern States, about as large as a dove, is our only species.

CUCKOOS.

The Cuckoos of the United States are about a foot long, with the upper parts of a metallic olive green color, and the under parts white. They are shy, concealing themselves in the thick foliage of trees, where they sit for hours uttering their unpleasant notes, which sound like *cow-cow*, eight or ten times repeated. They

Fig. 111.—Cuckoo.

feed upon insects, and also eggs, which they steal from the nests of other birds.

WOODPECKERS.

These birds have a straight, sharp bill, with which they cut into bark or wood in search of insects. The tongue is very long and capable of being greatly extended, and is armed towards the tip with barbs. By means of this instrument they pierce and drag forth insects from their hiding-places. Twenty or thirty kinds are found in North America, varying in size from the sparrow to that of a crow. They build their nests in holes, which they make with their bills in trunks or branches of trees.

Fig. 112.—Red-headed Woodpecker.

The Ivory-billed Woodpecker, of the Southern States, is the largest, and has the body black, with white upon the wings and neck, the crest scarlet, and the bill ivory white. The Black Woodcock, of the Northern States, is smaller, greenish-black in color, with a scarlet crest. The Hairy and Downy Woodpeckers, or Sapsuckers, are small, and black and white. The Red-headed Woodpecker has the head and neck crimson, the back, primaries, and tail black, the rump and a band on the wings white. The Golden-winged Woodpecker is larger

Fig. 113.—Golden-winged Woodpecker.

than a robin, and is one of our most beautiful birds. On the first sunny days of spring the Woodpeckers of this species appear on the tops of decayed trees, and as they hop about, striking with their bills here and there, make the woods resound with their loud, clear notes. Soon they pair, and both male and female begin to make a hole in a tree for the nest. The female lays from four to six beautiful white eggs for each brood, and two broods are reared in a season.

PERCHERS, OR INSESSORES.

These make up a large part of the most common birds, as Humming-Birds, Nighthawks, Kingfishers, Fly-catchers, Thrushes, Warblers, Creepers, Titmice, Sparrows, Grosbeaks, Larks, Blackbirds, Jays, Crows, &c.

HUMMING-BIRDS.

These are birds of the smallest size and of the most gorgeous plumage to be found in the feathered race. The beauty of their colors defies description; and from their brilliancy they are often called "flying gems."

Figs. 114 and 115. — Ruby-throated Humming-Bird and Nest.

There are about four hundred kinds, and they all belong to the continent and islands of America, and are most numerous in the warm regions. Their feet are very small, their wings long, and their power of flight very great; and they can balance themselves in the air, or beside a flower, with perfect ease. Their food consists of insects and the honey of flowers. Their nests are usually made of cotton, thistle-down, delicate fibres, and other soft materials, woven into a cup-shaped cradle, and placed on a branch of a tree not many feet from the ground; and the outside is covered with lichens in such a manner as to make the nest appear like a natural growth. The eggs are pure white.

WHIPPOORWILLS AND NIGHTHAWKS.

The Chuck-will's Widow, whose curious notes are heard in the evening and in the early morning in the

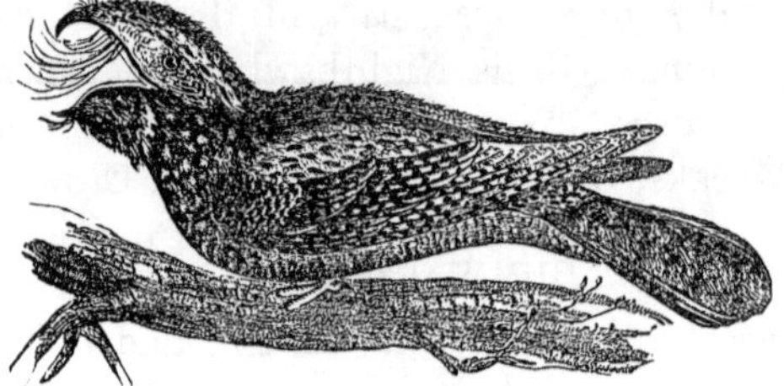

Fig. 116.—Whippoorwill.

Southern States, and the Whippoorwill and Nighthawk

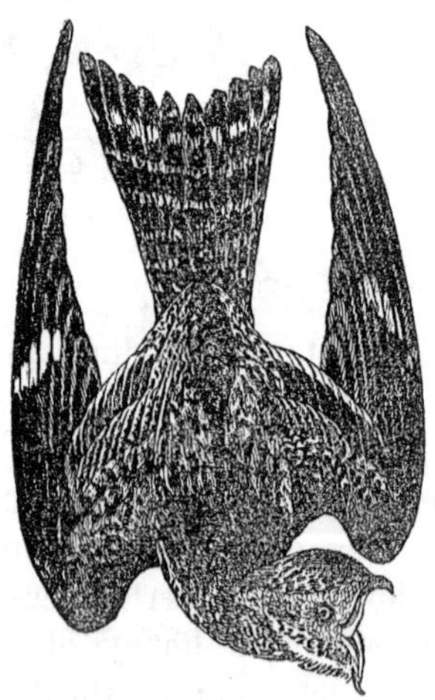

Fig. 117.—Nighthawk.

of the United States generally, are closely related to

each other. The last two are each about ten inches long, and of a dark color marked with white. The Chuck-will's Widow gets its name from its notes, which sound like the words *chuck-will's widow*, and the Whip-poorwill its name from a resemblance of its notes to the syllables *whip-poor-will*, which are also uttered in the evening and at early dawn. They make no nest, but lay their eggs on the ground, or on a flat rock.

KINGFISHERS.

These birds feed upon fish, and make their nests in holes which they dig in the banks of ponds and streams. They have a long, straight bill, and short legs. The Belted Kingfisher, of North America, is nearly as large

Fig. 118.—Belted Kingfisher.

as a small dove, the head crested, the color blue above and white below, with a blue belt. Sitting on a branch or decayed tree near the water, it watches intently for fish; and at the proper moment it plunges into the water, seizes its victim, flies to the nearest tree, swallows the fish, and is immediately on the lookout for another.

FLYCATCHERS.

There are about thirty kinds of these birds in North America. The bill is broad and bent down at the tip, and the sides of the mouth have stiff bristles. The Kingbird, Pewees, and Great-crested Flycatcher are some of the most common and best known species. The Kingbird is somewhat smaller than a robin, and is dark

Fig. 119.—Kingbird.

above and white below, with a hidden crest of orange, vermilion, and white. It is common in open fields and orchards, where it is seen perched upon a stake, tall weed, or low tree, watching for insects, which it darts down upon with sure aim. It is very courageous, eagerly attacking crows, hawks, and other large birds.

The Pewee, or Phœbe Bird, is smaller than the Kingbird, and its color is dark above and yellowish below.

It lingers around bridges, old buildings, and caves. Here in some secure spot it builds its nest of mud, grass, and moss, with a soft lining within for the eggs, which are pure white with reddish spots near the larger end. The Wood Pewee is rather smaller than the Phœbe, and is found in the quiet retreats of the forest.

THRUSHES.

The Wood Thrush, Hermit Thrush, Wilson's Thrush, Robin, Robin Redbreast, &c., come under this head.

The Wood Thrush is smaller than a robin, brownish above, white below, marked with triangular black spots.

Fig. 120. — Wood Thrush.

Fig. 121. — Ruby-crowned Wren.

It is found in groves and woods, and its sweet singing has made it celebrated among all lovers of birds. Its nest and eggs much resemble those of the robin.

The Hermit Thrush is smaller than the Wood Thrush, which it somewhat resembles, but it is rather darker above, and its breast is yellowish-white, and the dark spots beneath are less distinct than in the latter; and its soft, liquid, plaintive notes excel in sweetness those of any other American bird. It is heard in shady glens and deep woods.

The American Robin is one of the most common of the Thrushes, and its song in the early morning and at the close of the day is one of the pleasantest sounds that come from our groves and orchards.

The Robin Redbreast, of Europe, is about half as large as our robin, of a brown color, with a red breast. It loves to be near man, and often enters his dwelling. It is easily tamed, and is a great favorite. In severe weather it comes into the house, and, selecting a perch, warbles its song when the day is clear or when the fire burns brightly.

The American Bluebird is sky-blue above, and the breast chestnut-colored. Its nest is usually made in a hollow tree or post, and its eggs are from four to six, pale blue. It is a loving, gentle bird, and its soft warble is very pleasing. The Ruby-crowned Wren, which is now placed with the Thrushes, is scarcely more than four inches long, and is known by a patch of scarlet feathers on the crown. Its song is clear and sweet. The Water Ouzel, of the Rocky Mountains, is smaller

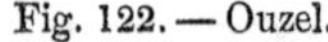

Fig. 122. — Ouzel.

Fig. 123. — Nightingale.

than the robin, and of a dark lead-color. This curious Thrush frequents mountain streams, into which it walks or dives, and moves about beneath the water in search of insects and other small animals upon which it feeds.

WARBLERS.

Warblers are among the smallest, most beautiful, and interesting of singing birds. Many kinds are generally found in the same locality, and may be seen gliding among the thick foliage, busily engaged in catching minute insects which hide beneath the leaves and in the buds and blossoms, and which often escape the sight of other and larger birds. Some of the Warblers are the sweetest of songsters, as the celebrated Nightingale, of Europe, shown in Figure 123. More than fifty kinds are found in the United States; and their very names are beautiful, and give us some idea of the appearance of these charming little creatures. Some of the more common are the Maryland Yellow-throat, the

Fig. 124. — Maryland Yellow-throat.

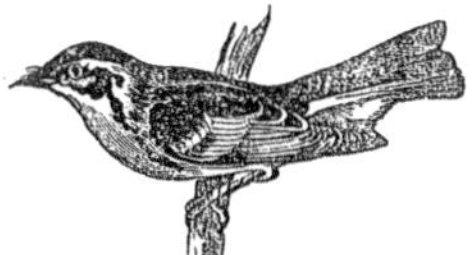

Fig. 125. — Blackburnian Warbler.

Blue-winged Yellow Warbler, the Golden-winged Warbler, the Orange-crowned Warbler, the Golden-crowned Warbler or Thrush, the Black-throated Green Warbler, the Yellow-rump Warbler, the Bay-breasted Warbler, the Chestnut-sided Warbler, the Blue Warbler, the Black Poll Warbler, the Yellow Warbler, the Black and Yellow Warbler, the Yellow Red Poll, the Yellow-throated Warbler, the Blackburnian Warbler, &c.

SWALLOWS.

These beautiful birds have long wings, short legs, and short, wide bill, and they spend much of their time upon the wing, skimming over fields and ponds, catching small insects, which constitute their food. One kind builds its nest upon the rafters in the barn, and is called the Barn Swallow; another builds its nest under the eaves, and is called the Eave Swallow; another under cliffs, and is called the Cliff Swallow; another digs a hole in a sandbank for its nest, and is called the Bank Swallow; and the Purple Martin comes and makes its nest in the Martin-houses which we place for it near our dwellings. Some persons suppose that these birds, which require air and sunshine as much as we do, spend the winter in the mud at the bottoms of ponds!

SHRIKES AND VIREOS.

The Shrike, or Butcher-Bird, is about as large as a robin, of a bluish color, with black wings and tail. Although belonging to the song-birds, it is a hawk in

Fig. 126. — Shrike, or Butcher-Bird.

its disposition, preying upon sparrows, warblers, and other small birds, as well as upon insects. It often imitates the cries of other birds, perhaps to call them from the trees and bushes, that it may get a chance to seize one of their number. It is called Butcher-Bird from its habit of impaling or hanging up its prey upon thorns and other sharp points, as a butcher hangs up his meats upon sharp hooks in his stall. It builds a large nest of twigs, grass, and moss, in the forks of a tree.

The Vireos are much smaller than the Shrike, and mostly olive-green above and light below. The Red-eyed Vireo has the iris of the eye red. Its loud, clear notes are heard in the tree-tops from spring till late in autumn. The White-eyed and the Warbling Vireo are small species, and their notes are very pleasant.

Fig. 127. — Warbling Vireo.

MOCKING-BIRDS, &c.

These birds are closely related to the Thrushes, and are very sweet singers. The Mocking-Bird, of the Southern States, is about the size of the robin, with a very long tail, and the color is ashy. It sings with great sweetness, and readily imitates the songs of all the birds which it hears. It is a very common pet in cages.

Fig. 128. — Mocking-Bird.

The Cat-Bird, of the Northern States, is smaller than the robin, and of a dark color, and in spring and the early part of summer its song is very mellow and sweet. Like its relative, it easily imitates the notes of other birds, and may be properly called the Mocking-Bird of the North. President Hill, of Harvard College, states that, having whistled a strain of Yankee Doodle two or three times in the presence of this bird, it imitated him perfectly. In the latter part of summer its notes are very harsh and disagreeable, sounding like the yawl of a cat.

Wrens are small birds, about the size of the Warblers. The Carolina Wren is one of the largest. It is reddish-brown. The House Wren delights in being near the habitations of man, and often makes its nest in a hole in the timbers or walls. The Winter Wren is one of the smallest, and of a brownish color. It is very active, and may be seen in twenty attitudes in the course of a minute.

Fig. 129. — Winter Wren.

CREEPERS, NUTHATCHES, AND CHICKADEES.

Creepers and Nuthatches are very small birds, which may be seen in North America at all seasons of the year, running along the trunks and branches of trees, and looking, at a little distance, much like little Woodpeckers. The American Creeper is light brown, with lighter streaks. The White-bellied Nuthatch is blue, with the under parts white, and the top of the head and neck black. The Red-bellied Nuthatch is a smaller species, and has the under parts red. Both kinds attach their feet to the bark, and sleep with their heads downward. The Chickadee is one of our smallest birds, and sings its simple *chickadee-dee-dee* in winter as well as in

Fig. 130.—American Creeper.

Fig. 131.—Chickadee, or Titmouse.

Fig. 132.—White-bellied Nuthatch.

summer, and in all sorts of weather. Its color is ashy above, whitish below, and the top of the head and throat black.

SKYLARKS.

The Skylark, or Shore Lark, is the only bird of its

Fig. 133.—American Skylark.

family in North America. It is smaller than the robin,

and sings sweetly while on the wing, but its song is short. The Skylark of Europe is almost as celebrated for its song as the Nightingale. It often rises vertically to a great height, and when rising or falling it sings its varied and powerful song.

FINCHES, CROSSBILLS, BUNTINGS, SPARROWS, AND GROSBEAKS.

The Purple Finch is about as large as the Bluebird, and of a beautiful crimson color; the female brown above and white below streaked with brown. The nest is built in a tree close to the ground, and the eggs are four, of a rich green color. The Yellow Bird, or Amer-

Fig. 134.—Purple Finch.

Fig. 135.—White-winged Crossbill.

ican Goldfinch, is of a beautiful yellow, the crown and wings black, tail and wings marked with white. The nest is very handsome, made of lichens, and fastened to a twig; eggs white, with a bluish tinge, and spotted with brown at the larger end.

Crossbills have the points of the bill much curved and crossing each other. By means of this curious instrument they can open the cones of pine and spruce with great facility, and thus secure the seeds, upon which they feed. Crossbills are about as large as the Bluebird; and there are two species in North America, — the Red Crossbill and the White-winged Crossbill, the latter having white bands upon the wings.

Fig. 136. — Song Sparrow.

Sparrows are plain-colored birds, generally dull brown, variously striped and marked, and are the most common in open fields, orchards, and about low bushes. There are many kinds in North America, all of which are small, the largest scarcely equalling the common Bluebird in size. Some of the principal kinds are the Bay-winged Bunting, the Yellow-winged Sparrow, the White-crowned Sparrow, the White-throated Sparrow, the Black Snow-Bird, the Tree Sparrow, the Chipping Sparrow, the Song Sparrow, the Swamp Sparrow, the Fox-colored Sparrow, &c.

Fig. 137. — Rose-breasted Grosbeak.

The Grosbeaks have the bill very large, and hence their name, which means *great beak*. The Rose-breasted Grosbeak is one of the most beautiful of the North American birds. It is smaller than a robin, and the color is black and white, the breast a rich carmine. The female has no black or carmine. The song is loud, clear, and sweet.

The Ground Robin, Towhe Bunting, or Chewink, is about two thirds as large as a robin, the color black and white. The female is brown and white. It is seen almost everywhere, in low bushes, in fields, or by the wayside, and is easily found out by its sweet *chewink*, which it utters every few moments. Often near the close of day in spring it mounst the topmost twig of a small tree, and sings with a sweetness that charms all who listen to it. It makes its nest upon the ground, and lays from four to six eggs of a light color with dark spots.

Fig. 138. — Chewink.

BLACKBIRDS, LARKS, &c.

The Bobolink, Cow-Bird, Blackbirds, Larks, and Orioles belong to one family. The Bobolink is somewhat larger than a Bluebird, of a black and cream color, the female yellowish brown. Its jingling song, uttered from a low tree, or bush, or tall weed, or upon the wing, is familiar to all who live in the country. Late in the summer Bobolinks fly southward, and are seen in immense flocks in grain-fields and along the margins of creeks and rivers, where the tops of the reeds are bent with ripe seeds. Thousands are shot by the hunters and sold in the markets, where they are called Reed-Birds.

Fig. 139. — Bobolink, or Reed-Bird.

The Cow-Bird is larger than the Bobolink, and is the most singular bird in North America. For some reason which is not understood it never makes a nest, but, like the European Cuckoo, stealthily lays its eggs, only one in a place, in the nests of Warblers, Flycatchers, Bluebirds, Sparrows, and the Golden-crowned Thrush. The egg is grayish blue, marked with brown dots and short streaks. And it is a curious fact that this egg hatches before the eggs of the bird in whose nest it is laid. As soon as the young Cow-Bird is hatched, the foster-parents leave their own eggs to get food for it, and hence the young in their eggs die, and the eggs are soon thrown from the nest. Then the young Cow-Bird receives the whole attention of those that have been compelled to adopt it, and they feed it till long after it can fly, and until it is larger than the foster-parents themselves. The head and neck of the Cow-Bird is of a chocolate color, the rest of the body lustrous black; the female is light brown.

The Red-winged Blackbird is nearly as large as the robin, shining black, with the shoulder and a part of the wing bright crimson. The female is of a dusky color. It is common about ponds and marshes, and builds its nest in low bushes or tufts of sedges.

Fig. 140. — Meadow Lark.

The Meadow Lark is rather larger than the robin; the upper parts brown and brownish white, the under parts yellow, with a black crescent upon the breast. The nest is built at the foot of a tuft of grass, and is covered over, except the entrance.

The Baltimore Oriole, or Golden Robin, is as large as a sparrow, the color black and orange-red, and is one of the most beautiful birds in the United States. Its song is loud, full, and mellow. Its hanging nest is woven to the outer-drooping twigs of the elm and other trees. It is often made of fibres from the silkweed.

CROWS, RAVENS, JAYS, AND MAGPIES.

These are rather large birds. The Raven is the largest. It is but seldom seen east of the Mississippi. The Crow is well known, and farmers regard it as their enemy, because it pulls up the young corn; but it does much more good than harm, by destroying a great number of grubs, which would injure the crops. The Blue-Jay is a bird of wonderful beauty, but its notes are harsh, and it eats the eggs of other birds, and even destroys young birds, swallowing them greedily. The Magpie is about as large as a dove, black and white, and the tail is very long. There are two kinds in North America, and one in Europe. It is a noisy bird, and it can be taught to speak.

Fig. 141. — Magpie.

SCRATCHERS, OR RASORES.

Doves, Wild Pigeons, Turkeys, Hens, Grouse, Pheasants, and Quails are the principal Rasores. Most of them live mainly upon the ground, and all feed upon seeds, grain, nuts, and berries. The Rasores are very important to man, their flesh furnishing him with some of his choicest food. Excepting the Doves and Pigeons, they are able to run as soon as hatched.

PIGEONS.

The Wild Pigeon, of North America, is about as large as a dove, with a very long tail, and the color above is

Fig. 142. — Wild Pigeon.

blue, under parts reddish, and the neck glossy golden violet. It flies very rapidly, and sometimes millions move together, darkening the air like a cloud, and, on alighting, fill forests, and even break down large trees by their weight.

GROUSE.

The Prairie Chicken, Ruffed Grouse, Ptarmigans, &c. come under this head.

The Prairie Chicken is about as large as a common hen, and the male has an air-sack on each side of the neck by which it is able to produce a loud booming sound. The Ruffed Grouse, or Partridge, of the United States is rather smaller than the common hen, and has a beautifully barred and spotted plumage. This bird prefers open woods and the borders of forests, and in

Fig. 142 a. —Ruffed Grouse.

winter thickets of evergreens. When disturbed it takes wing with a loud whir. In the spring the male, while standing upon an old log, makes a loud sound with his wings, which is called drumming. The female makes her nest of leaves upon the ground, and lays a dozen or more dingy-white eggs.

QUAILS.

These birds are much smaller than the Grouse, and about forty kinds are found in America. The Quail

Fig. 143. — Quail.

has a body about as large as a pigeon, and its color is reddish brown. In Pennsylvania and southward it is

Fig. 144. — Mountain Quail.

called the Partridge. Its notes are a sort of whistle. The nest is built near a tuft of grass, and the eggs are from ten to eighteen, pure white.

The Mountain Quail is found in Oregon and California.

RUNNERS, OR CURSORES.

These are the Ostriches and their relations. They are very large birds with long legs and rudimentary wings. The Camel-Bird, or great Ostrich of the deserts of Africa and Asia, is about eight feet high, and has only two toes to each foot. The Rhea is a three-toed Ostrich of South America. The Cassowaries are three-toed Ostriches which inhabit the Indian Archipelago and Australia. And the Apteryx is an ostrich-like bird of New Zealand. The Great Bustard of Europe, a bird which attains a weight of thirty pounds, also belongs in this group; its wings, however, are not rudimentary, though very short.

WADERS, OR GRALLATORES.

The Waders have a long bill, long neck, and long legs. They are the Cranes, Herons, Ibises, Plovers, Turnstones, Stilts, Woodcocks, Snipes, Yellow-Legs, Godwits, Curlews, Rails, and Gallinules. They live mainly in wet places, or upon marshes or shores, and are adapted by their long legs for wading in shallow waters. They feed upon worms, shell-fish, and other aquatic animals. Figures 145–157 show some of the common kinds.

HERONS.

The Great Blue Heron, of North America, frequents ponds and creeks, where it may be seen standing for hours, upon a rock or stump, watching for fish. When wounded it is dangerous to approach it, as it strikes with its bill, and generally aims at the eye. This

Fig. 145.—Great Blue Heron.

Heron is three feet and a half long. It builds its nest on a large tree, in a dense swamp.

The Bittern, or Stake-Driver, and the Night Heron,

Fig. 146.—Bittern, or Stake-Driver.

with its long, white plumes, are much smaller species.

IBISES.

The Wood Ibis is nearly as large as the Great Blue

Fig. 147.—Wood Ibis.

Fig. 148.—Plover.

Fig. 149.—Turnstone.

Fig. 150.—Yellow-Legs.

Fig. 151.—American Woodcock.

Fig. 152.—Wilson's Snipe.

Fig. 153.—Stilt.

Fig. 154.—Godwit.

Fig. 155.—Curlew.

Fig. 156.—Rail.

Fig. 157.—Gallinule.

Heron, and lives in the swamps of the Southern States. In order to obtain food, it moves about in the shallow waters until these become muddy, when the fishes rise to the surface, and are struck and killed by its bill.

SWIMMERS, OR NATATORES.

These birds are fitted to live in and about the water. Their feet are webbed, and the plumage is thick and made water-proof by the oil with which they dress it. They swim with great ease, and most of them are expert divers. Swans, Geese, Ducks, Pelicans, Petrels, Gulls, Divers, Auks, and their relatives, belong in this group.

SWANS, GEESE, AND DUCKS.

The Swans have the neck very long, and they are much larger than the largest goose. There are two species in North America, — the American Swan and the Trumpeter, both pure white.

The Wild Goose is larger than the common goose, of a brownish color, with black head, neck, bill, feet, and tail. Wild Geese are seen in early spring in large flocks, moving northward, where they rear their young, returning south in autumn. The peculiar noise made by a flock as they pass over is familiar to all. They are sometimes tamed, but often manifest a desire to join the migrating flocks. Wilson says a wild goose was captured by a farmer on Long Island, and kept all winter with a flock of common geese. The following spring it joined a party of its own kind which was passing over. The next autumn, as a flock of wild geese was returning southward, and passing directly over this man's farm, three of the number separated from the others and alighted in the yard. They proved to be the long-lost goose and two of her young.

The Mallard, or Greenhead, is about two feet long, and has the plumage of the head bright green; there is a white ring around the neck, and the general color of the body is brownish. This is the parent of the domestic duck.

The Wood Duck is smaller than the Greenhead, and its plumage excels in beauty that of all other ducks.

Fig. 158. – Wood Duck.

It builds its nest in a hollow tree or limb; and if the nest is over water, the young, as soon as hatched, drop into it; if not, they fall to the ground, and are led or carried to the water by the parent.

The Canvas-Back is about the size of the Wood Duck, with a chestnut-colored head, and the other parts white and black.

The Eider Duck is one of the largest of the Ducks; colors black and white. It lives in the cold regions of the North. Eider-down is obtained from the nests of

Fig. 159. — Canvas-Back.

Eider Ducks, the birds having plucked it from their breasts to place around their eggs.

ALBATROSSES AND PETRELS.

The Albatrosses are the largest of web-footed birds;

Fig. 160. — Sooty Albatross.

the Petrels, in many cases, are very small. Both live on the ocean, but come on shore to rear their young. The voice of the Albatross is as loud as that of an Ass.

The Stormy Petrels, or "Mother Carey's Chickens," are the smallest of web-footed birds; but they are able to fly about during the most terrific storms. While flying close to the water they extend their legs, and thus appear to walk upon its surface. The word *Petrel* means *little Peter*.

Fig. 161.—Stormy Petrel.

GULLS AND TERNS.

The Gulls and Terns have long and pointed wings, and are common upon the shores of all countries, and also on the larger rivers and lakes. They swim well, but do not dive. The Gulls are generally light-colored,

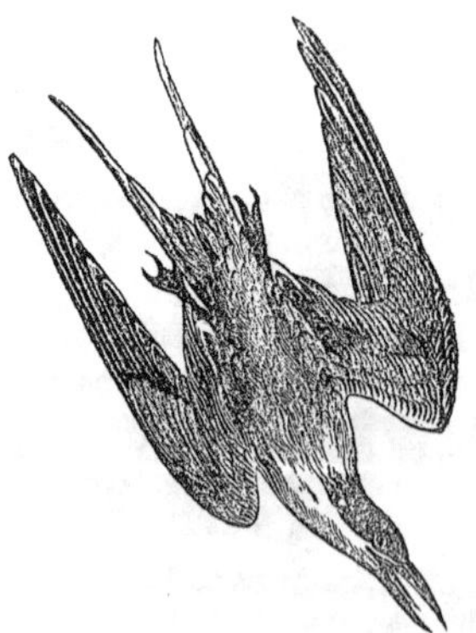

Fig. 162.—Tern.

and they vary in size from that of a dove to that of a goose. The Terns have the tail very long and forked. They are generally light below, black, and bluish above. They are of the size of a dove and smaller, some being no larger than a robin. They feed upon small marine animals.

DIVERS.

The Great Northern Diver, or Loon, is almost as large

Fig. 163. — Great Northern Diver, or Loon.

Fig. 164. — Crested Grebe.

as a goose, black above, beautifully spotted with white, and white below. It is exceedingly keen-sighted and wary, and it dives so quickly that, seeing the flash of the gun, it is often under water before the shot reaches it.

Grebes are divers which are smaller than the Loon, and in the spring have the head ornamented with tufts of feathers. When alarmed, they remain beneath the surface of the water, exposing only the bill.

AUKS, PUFFINS, AND PENGUINS.

These belong to the cold regions, and the Penguins to the Southern Hemisphere. The Great Penguin of Patagonia is larger than a goose. Its wings are so

Fig. 165.—Patagonian Penguin.

small that it cannot fly, and it stays in the water, except when it crawls on shore to lay its eggs and rear its young. Penguins hatch their eggs by holding them between their thighs, and move away with them if disturbed when sitting.

Fig. 166.—Puffin.

The Great Auk of the Arctic regions is as large as the Penguin. Other kinds are much smaller, and those called Puffins are not larger than a dove. Puffins make their nests in burrows in the ground, and each bird lays but one egg in a season.

REPTILES.

Reptiles are vertebrates which have cold blood, and are covered with hard plates, called scales, and which lay eggs in holes that they dig in the ground; these eggs hatch without being brooded by the parent, and the young, as soon as hatched, look just like the parents, only smaller. Reptiles are such as Turtles or Tortoises, Lizards, and Serpents or Snakes.

TURTLES.

Turtles, or Tortoises, are reptiles which have a shell into which they can more or less completely withdraw their head, legs, and tail. Some of them live wholly upon the land, like those called Gophers in the Southern States, which dig burrows in the ground that are dangerous pitfalls for horsemen, and the Box Turtles, which live in the woods, and which can shut their shell so tightly as to entirely hide their extremities, as seen in Figure 169. Others, like the Painted Turtle, with its colors of black, yellow, and red, the Wood Tortoise, with its beautifully carved scales, the Speckled Tortoise, with its black shell ornamented with orange-colored dots, and the Snapping Turtle, live in fresh-water ponds and streams, coming at times upon the land. Others, like the Salt-water Terrapin, so much prized for food, live in salt-water creeks. Others, like the Hawk-bill Turtle, the Green Turtles, and the Soft-shelled Sphargis, live in the ocean, and only come on shore to lay their eggs. The land and fresh water turtles of North America have the shell from four to six or eight inches long; excepting the Gophers and Snappers, which are much larger, having the shell a foot and a half or more in length, and, in some cases, the Snapping Turtle is four feet long from the nose to the tip of the tail. This turtle has the head and neck very large, and the jaws strongly hooked, and it is exceedingly powerful, and very voracious, devouring smaller reptiles, fishes, young ducks, and other small animals. When molested it raises itself on its legs, opens its mouth wide, and, throwing the body forward, snaps its jaws upon its enemy with fearful power. See Figure 170.

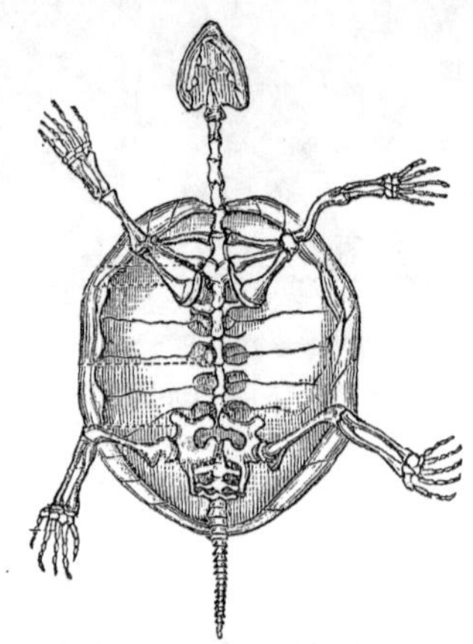

Fig. 167. – Skeleton of a Turtle.

The Hawk-bill Turtle, Figure 171, lives in the warm parts of the Atlantic Ocean, and weighs about two hundred pounds, and its scales furnish the material for the beautiful and costly tortoise-shell ornaments.

The Green Turtles weigh two or three hundred

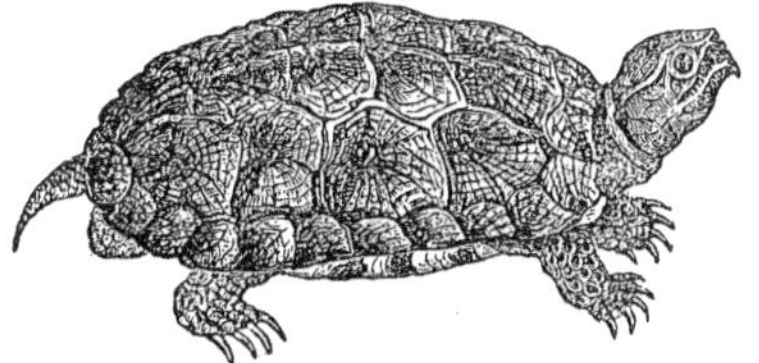

Fig. 168.—Wood Tortoise.

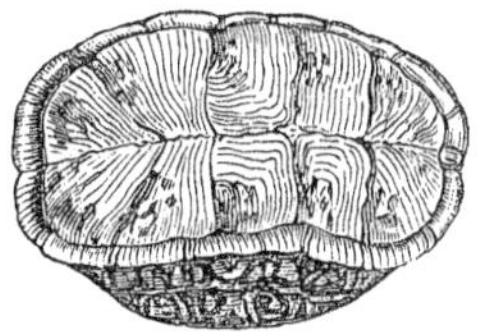

Fig. 169.—Box Turtle, shut up and on its back.

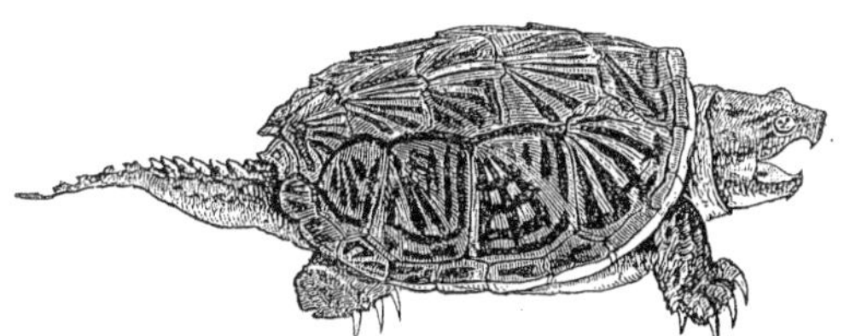

Fig. 170.—Snapping Turtle.

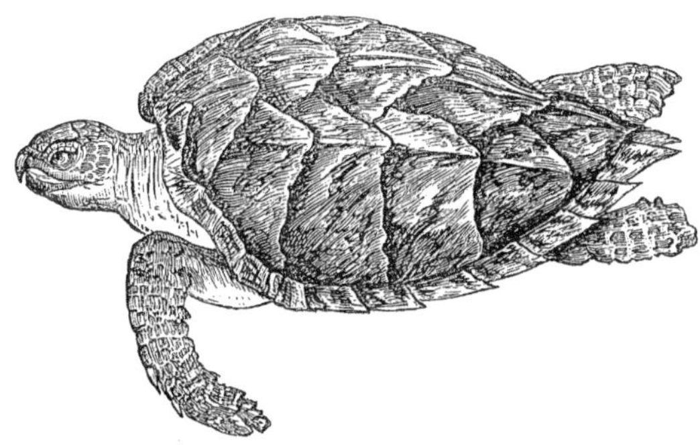

Fig. 171.—Hawk-bill Turtle.

pounds, or more, and are caught at night when they come on shore to lay their eggs.

The Sphargis, or Soft-shelled Sea Turtle, lives in the Atlantic Ocean and in the Mediterranean Sea, and is the largest of all the turtles, sometimes weighing two thousand pounds! It is covered with a thick leather-like skin instead of a hard shell.

LIZARDS, OR SAURIANS.

Lizards have a long body, long tail, no shell, and the mouth is large and armed with teeth. Enormous lizards, thirty feet long, live in the river Nile, and are called Crocodiles. The Alligators, of the Southern

Fig. 172. — Alligator.

States, are lizards which are five, ten, or fifteen feet long, and which have a head shaped something like that of a pickerel. They are numerous in the creeks and sluggish streams, and devour all kinds of small animals which come in their way.

The Six-lined Lizard, of the Southern States, is only nine or ten inches long, with six yellow lines along its sides and back. It is harmless, runs rapidly, and feeds upon insects. The Green Lizard, of the Southern States, is a smaller species which is common about gardens and buildings, often entering houses, and moving over the furniture, up and down the walls and

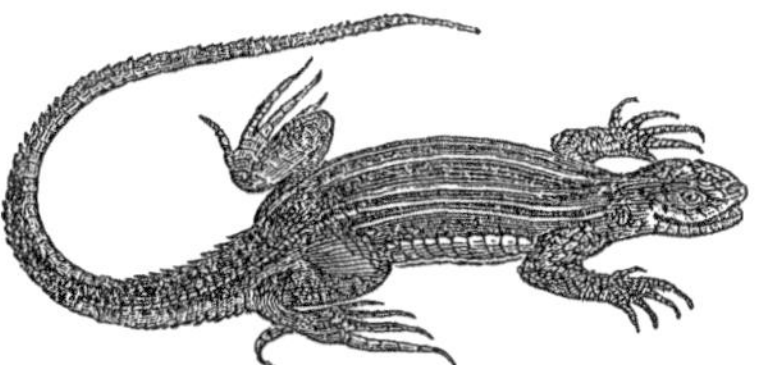

Fig. 173.— Six-lined Lizard.

window-panes, and along the ceilings, in its search for flies, upon which it likes to feed.

The Horned Toads are Lizards found in the southern and western parts of North America. They have the head armed with spines, and the body covered with tubercles. The Horned Toad, of Texas, is less than five inches long, and is lively in its movements. It is sluggish when kept in a cage.

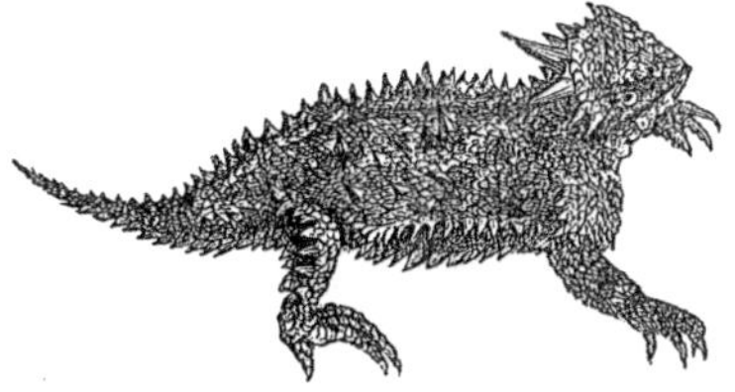

Fig. 174.— Horned Toad.

SERPENTS, OR SNAKES.

Serpents are reptiles which are exceedingly long in proportion to their size, and which have no feet, yet they glide over the ground with very great speed. Their mouth, throat, and body are capable of being greatly distended, and hence they are able to swallow animals whose bodies are much greater in diameter than their own. They do not masticate their food, and hence their teeth are suited only for seizing, killing, and retaining prey. The tongue is long, and capable of being run out much beyond the mouth, and it can be concealed within a sheath at its roots. They shed their skins every year, and most of them lay eggs from which the young are hatched. There are more than a thousand kinds of snakes, and more than a hundred kinds in North America. Some of the largest

Fig. 175. — Black Snake.

in the Tropical regions, as the Boas and Anacondas of South America, and the Pythons of Africa and India, are thirty or forty feet long, and are able to swallow dogs, deer, and even oxen after they have crushed them in their powerful folds.

The Black Snake and the Striped Snakes are the most common kinds in North America. The former is from three to five feet long, and lustrous black. It runs very fast, and climbs trees and bushes with great ease, where it seeks bird's nests and devours the young. It is feared by most persons, but it is harmless.

The Rattlesnake, of North America, is found on rocky hills and mountains, and its bite is almost always fatal to men and animals. It has two very sharp fangs in the upper jaw, and these fangs are hollow or grooved, and connected with a bag of poison, so that when the snake strikes them into an animal, the poison is forced into the wound.

BATRACHIANS, OR FROGS, TOADS, SALAMANDERS, &c.

These are reptiles which have no scales, and which lay their eggs in the water, and the young resemble

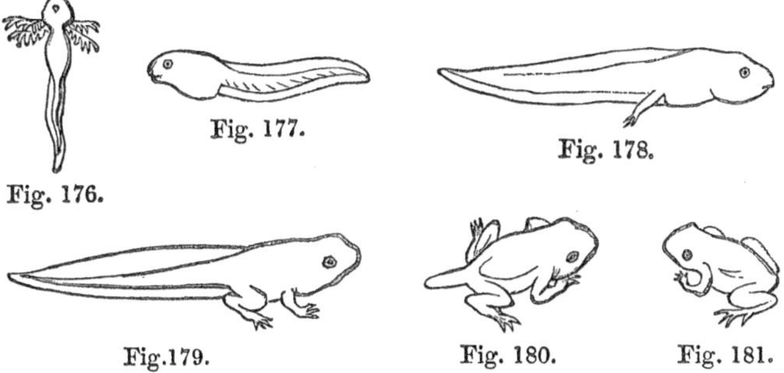

Figs. 176 - 181.—Changes in the form of a Frog from the time of Hatching.

fishes more than they do their parents. The young breathe by means of gills, like fishes, but the adults breathe by lungs, like those reptiles which have scales. For example, the young frog or tadpole, when first hatched, appears as in Figure 176, with the gills in tufts on the two sides of the neck; later, it appears as in Figure 177, where the gills are concealed; later, it appears as in Figure 178, where it has hind legs; later, as in Figure 179, with four legs; later still, as in Figure 180, where the tail has mostly disappeared; and later still, it becomes a perfect frog.

FROGS AND TOADS.

These have the body short and thick, and the tongue is long and fixed to the fore part of the jaw, and its tip is turned backward into the mouth, from which it can be darted forth quicker than a glance of the eye; and it is by means of the tongue that frogs and toads snap up insects and worms, which form their principal food. The Bull-frog is our largest kind, and is well known by its croakings, which may be heard a mile. The Green

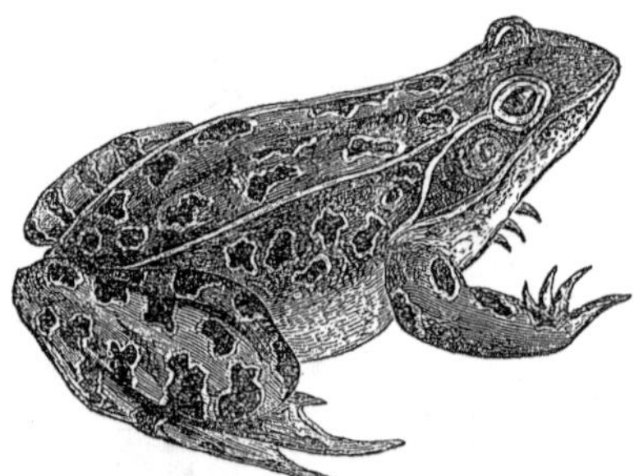

Fig. 182.—Leopard Frog.

Frog, Leopard Frog, Pickerel Frog, are other kinds that are found about ponds and streams. The Wood-frog is found in the woods, and goes to the water only in spring, when it lays its eggs. The Tree-frogs, often called Tree-toads, have the toes so formed that they are able to move along the trunks, branches, and leaves of trees. Here they live, except when they go into the water to lay their eggs. One of the tiny Tree-frogs, named Pickering's Hylodes, makes the high piping note, which in spring is heard in New England and in the Middle States throughout the night. It is found upon plants near to stagnant pools, and in woods.

Fig. 183. — Pickering's Hylodes.

The American Toad is familiar to all. It is very useful to the farmer and gardener, destroying great numbers of insects.

SALAMANDERS, TRITONS, SIRENS, &c.

Salamanders are batrachians which have a long body and long tail, and which live upon the land, except when they go to the water to lay their eggs. There

Fig. 184. — Salamander.

are many kinds in North America, varying from three to twelve inches long. They are found mostly under

Fig. 185. — Triton.

stones, fallen trees, and rubbish. Tritons have nearly the same form, but live in the water. Both correspond in form to the Lizards among the true Reptiles. Tritons have the most wonderful power to repair or renew injured or lost parts. The legs may be cut off, and in less than a year they will grow again; and the limbs thus formed may also be cut off. and others will grow in their places; and even if the eye be destroyed another will grow to supply the loss! In the Southern States is found the Congo Snake, an animal which is related to the Salamanders and Tritons. It is about two feet long, and lives in muddy waters. The Sirens have the gills in tufts, as in Figures 187, 188; thus even in the adult state they are like the young of Frogs and Toads.

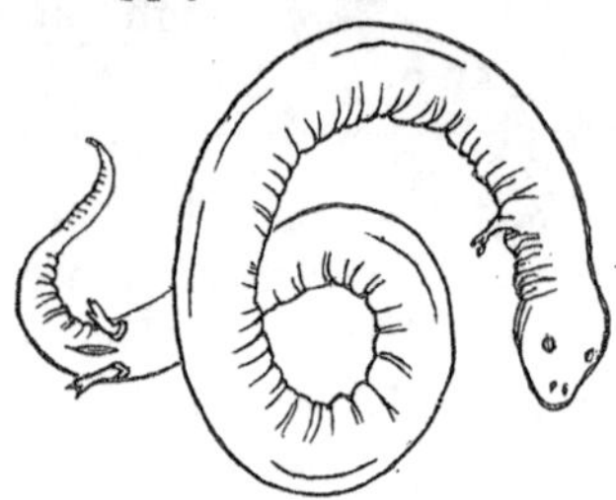
Fig. 186. — Congo Snake.

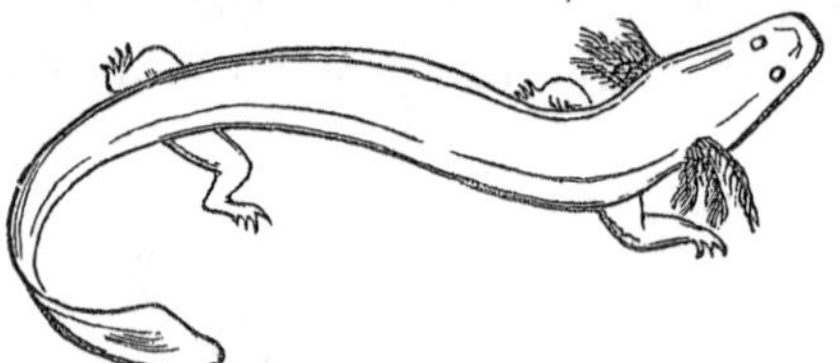
Fig. 187. — Menobranchus, or Mud Puppy.

They live in the water. The Menobranchus, or Mud

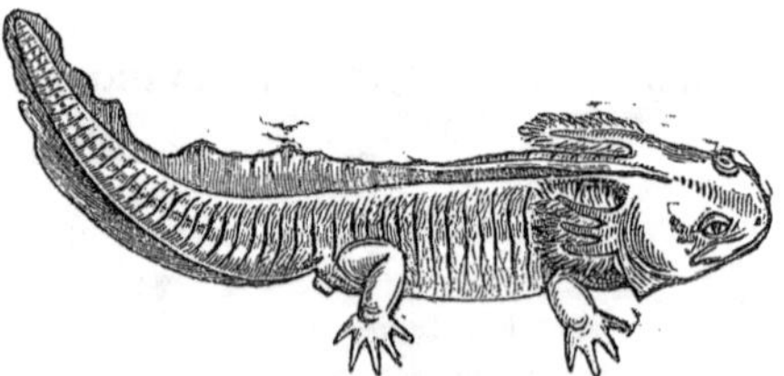
Fig. 188. — Axolotl.

Puppy, of our Northern Lakes, the Siren, of the Southern States, and the Axolotl, of Mexico, are of this kind. The two last are six to twelve inches long.

FISHES.

Fishes are vertebrates which have cool blood, and live wholly in the water, and breathe by means of gills. Most of them are scaly, but some are covered with a smooth skin, others have spines, and others still are covered with bony plates. The jaws are generally armed with teeth, and, in many cases, all parts of the mouth also, and even the gullet. Their movements are generally rapid, and their forward motion is mainly produced by the movements of the tail. The parts which correspond to the arms and legs of quadrupeds are very short, and are called fins; and their use is mainly to balance and direct. The flesh is light-colored or white. In general, the eye of Fishes has no motion, and the pupil is always of the same size, both in light and darkness; and the ear is wholly enclosed by the bones of the head, and hence it is generally believed that they hear only the loudest sounds. They are very voracious, feeding mainly upon smaller fishes, and other small animals, which they usually swallow whole. Those which feed on shell-fish crush their food by means of the teeth in the gullet. Most fishes lay eggs; a few kinds bring forth living young. Nearly all seem to have no care for their young, but eat them as greedily as they do other food. The number of eggs laid by a single fish in one season is often very great, — the Salmon sometimes laying twenty thousand, and the Cod more than nine millions! The colors of Fishes are very beautiful, exhibiting a lustre like that of the metals, and the brilliancy of precious stones, and the delicate tints of flowers; they are indeed the gems of the waters, as the humming-birds are the gems of the air. The wonder-

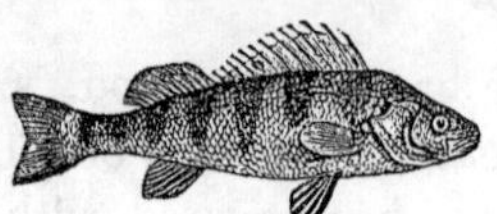

Fig. 189. — Yellow Perch.

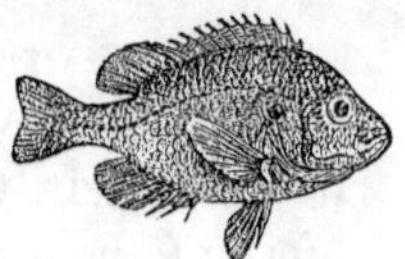

Fig. 190. — Bream.

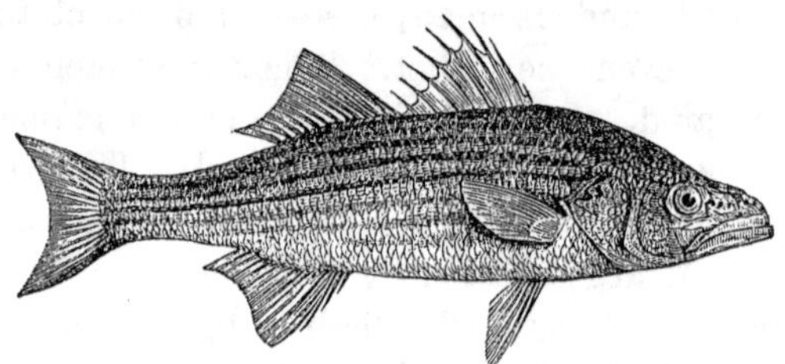

Fig. 191. — Striped Bass.

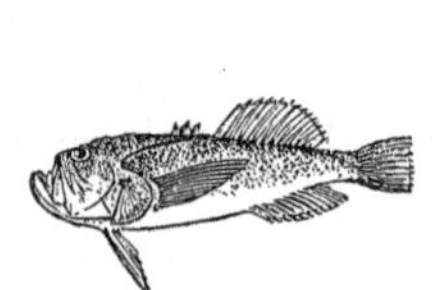

Fig. 192. — Star-Gazer.

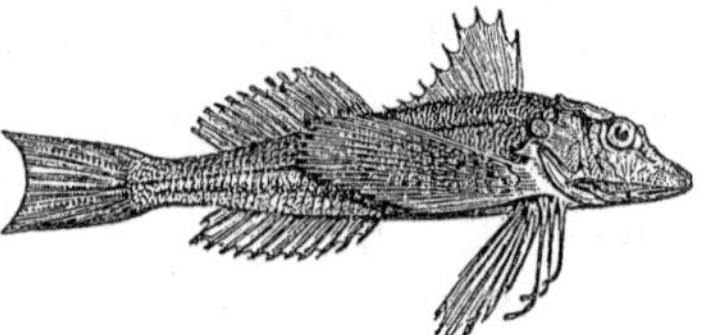

Fig. 193. — Sea Robin.

Fig. 194. — Stickleback.

Fig. 195. — Darter.

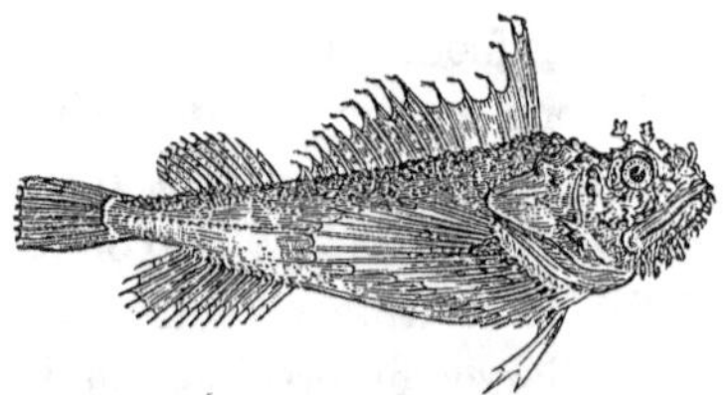

Fig. 196. — Sea-Raven.

ful power and swift motion of some, the wholesome and delicious food furnished by many, and the exciting sport of their capture, combine to render Fishes objects of great interest to almost every one. The number of known kinds is about ten thousand.

SPINE-FINNED FISHES.

Spine-finned Fishes have spines in the back or dorsal fin, and often in the lower fins. The Perch, Sea-Bass, Pond-Fish or Bream, Star-Gazers, Sculpins, Sticklebacks, Porgees, Mackerel, Sword-Fishes, and a host of others belong to this group; for it is the largest of all the groups of Fishes.

The American Yellow Perch, of our ponds and rivers, is known to every boy. The Striped Bass is caught in the sea near the shore, and the largest weigh seventy-five pounds each. The Pond-Fish or Bream is found in every pond, and the round cavities, which it makes for its nest, may be seen in great numbers near the shore. The Star-Gazers live in the sea, and have the eyes on top of the head, so that they appear as though looking at the heavens. The Sculpins live in the sea, and are often called Sea Robins, Sea Ravens, &c. The Sticklebacks are very small fishes which inhabit both the sea and streams, and are very active and greedy, a single one having devoured seventy-five young fish in less than half a day! They construct very curious nests. The Weak Fish and Porgees live in the Atlantic Ocean, and are caught for food. The Mackerel lives in the sea, and is caught on the coast of New England in immense numbers. The Sword Fish has the upper jaw very much extended, forming a powerful and dan-

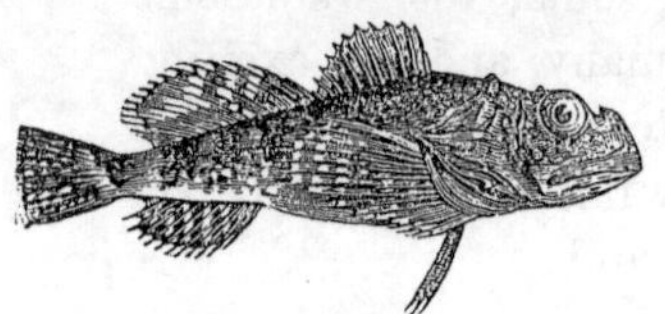

Fig. 197.—Sculpin.

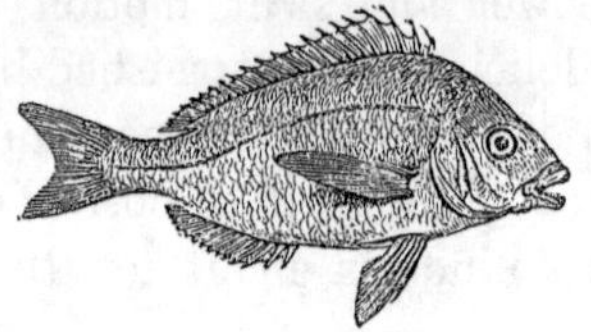

Fig. 198.—Scupaug.

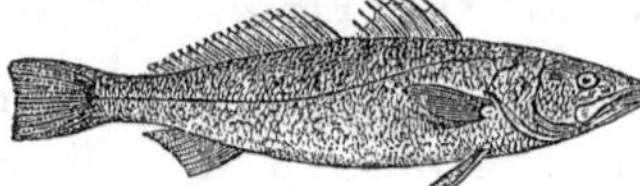

Fig. 199.—Weak Fish.

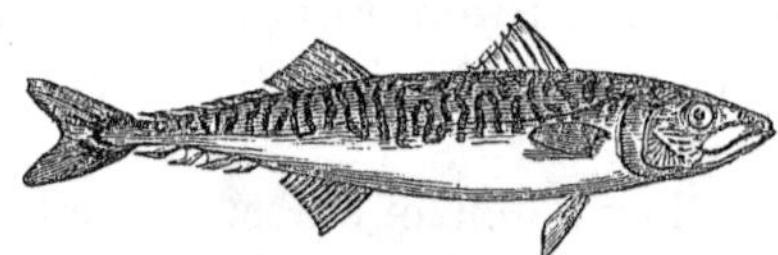

Fig. 200.—Mackerel.

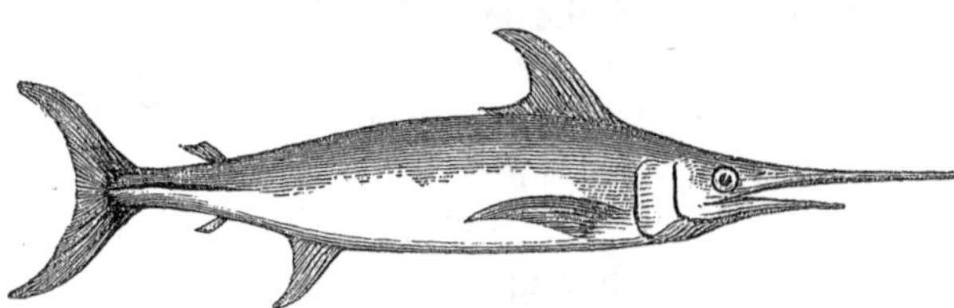

Fig. 201.—Sword Fish.

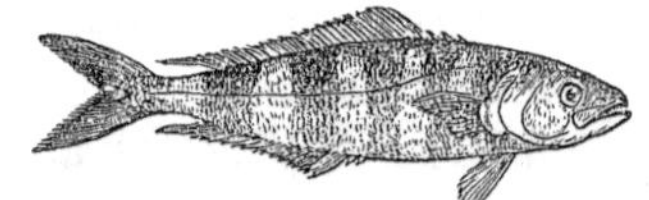

Fig. 202.—Pilot Fish.

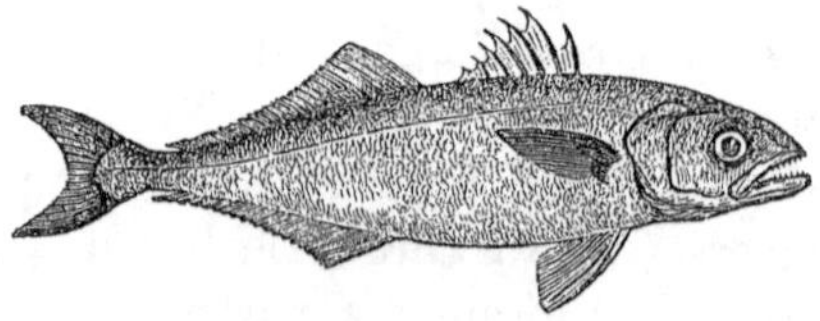

Fig. 203.—Blue Fish.

gerous weapon, with which it attacks Whales and other large animals of the sea. The Blue Fish is found in nearly all seas, and makes excellent food. The Dolphin lives in the Mediterranean, and in the Atlantic, and is celebrated for its beautiful colors, and for the brilliant tints which it has when dying. The Surgeon has a sharp spine or lancet on the side of its tail; it lives in the sea. Mullets are small fishes which live in the sea, and in fresh waters. Eel-Pouts are long, somewhat eel-shaped fishes, which the fishermen catch when fishing for cod. The Goose Fish, of the Atlantic, is large, sometimes weighing seventy pounds, and with such a big mouth that it swallows fishes almost as large as itself. Gulls and other sea-birds are often found whole in its stomach! The Toad Fish, of the Atlantic, is about a

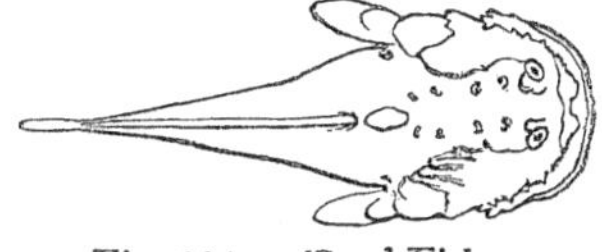

Fig. 204.—Toad Fish.

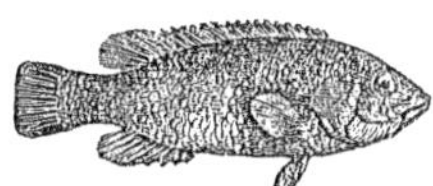

Fig. 205.—Conner.

foot long, and seems to care for its young. The Conner is very abundant on the coast of New England, and is caught in great numbers both with the hook and net.

SOFT-FINNED FISHES.

These Fishes have no spines in their fins. They are the Carp, Dace, Shiners, Suckers, Pike, and Pickerel, Gar-Fishes of the Sea, Flying-Fishes, Salmon, Herring, Cod, Eels, &c.

The Common Shiner, found in most ponds, lakes, and rivers, is from three to six inches long, and of a golden color. The Pickerel, so well known in the fresh waters,

Fig. 206. — Blunt-nosed Shiner.

Fig. 208. — Surgeon.

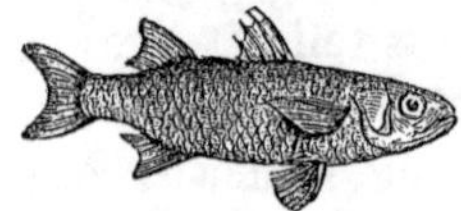

Fig. 207. — Mullet.

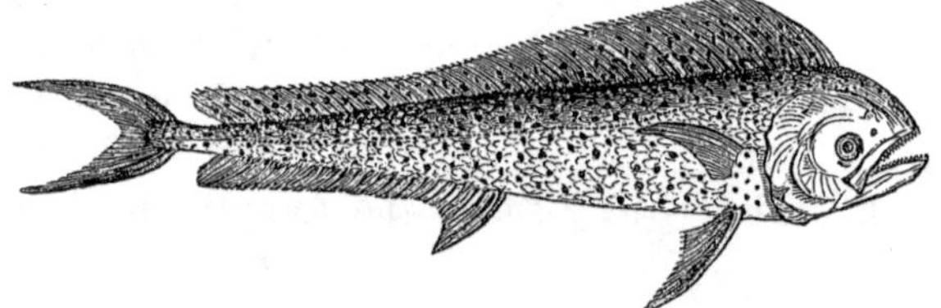

Fig. 209. — Dolphin.

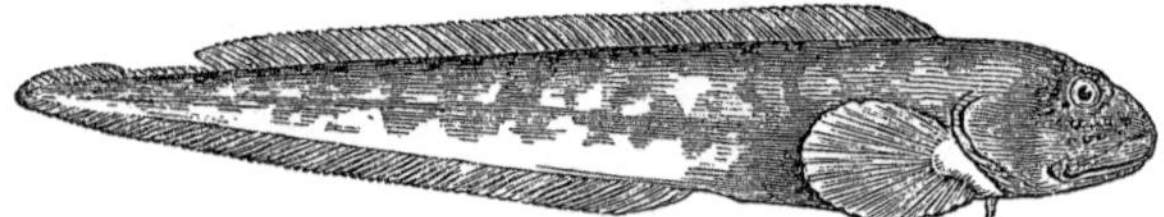

Fig. 210. — Eel-Pout.

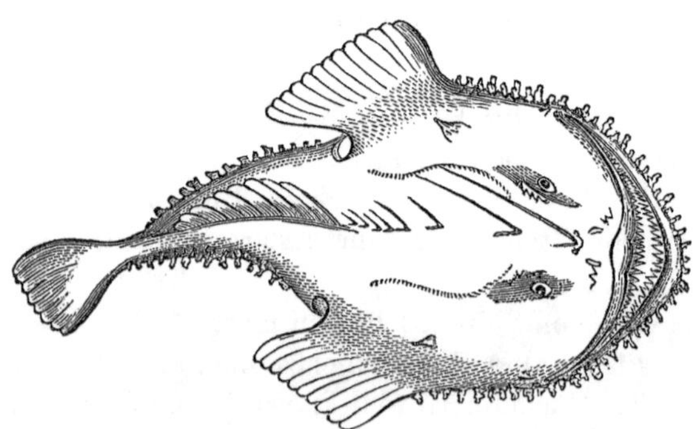

Fig. 211. — Angler, or Goose Fish.

is a handsome fish, of fine flavor, and the sport of its capture is very exciting. The Gar-Fish lives in the sea, and has an extremely long head and body; the jaws are pointed, and armed with many small teeth, and its bones are green. Flying Fishes have the fins, which are directly behind the gills, so large that they are able to sustain themselves in the air for a few moments, thus appearing to fly. They live in all warm and temperate seas, and are from three inches to a foot in length. The Blind Fish is found in the waters of the Mammoth Cave, Kentucky, and is about three inches long. Its eyes are under the skin, so that the fish is perfectly blind, and thus adapted to the dark waters of the cave. The Horned Pout, from six to ten inches long, and common in ponds and sluggish streams, has the head armed with sharp spines, which inflict a smarting wound on the hand of the careless fisherman. The Salmon is a most beautiful fish, whose home is in the Arctic seas, but it comes southward and ascends rivers for the purpose of laying its eggs, and is caught in large numbers. Its flesh is delicious, and it weighs from ten to thirty pounds or more. The Lake Trout inhabits our Northern lakes, and is from two to five feet long, of a gray color with lighter spots. It is sometimes called the Longe. The Brook, or Speckled Trout, is found in most of the clear streams of the temperate parts of North America, and is very beautiful, being dark above, silvery below, and the sides dotted with red and yellow. Its flesh is delicate. It is very shy, and its capture often requires much skill. The Herring lives in the Arctic seas, and comes southward in spring to lay its eggs. It is about a foot long. The Gar-Pike has a long body and long jaws, which have

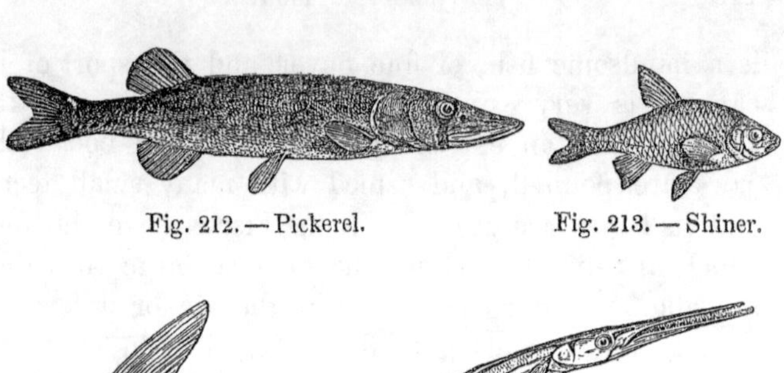

Fig. 212. — Pickerel.

Fig. 213. — Shiner.

Fig. 214. — Flying Fish.

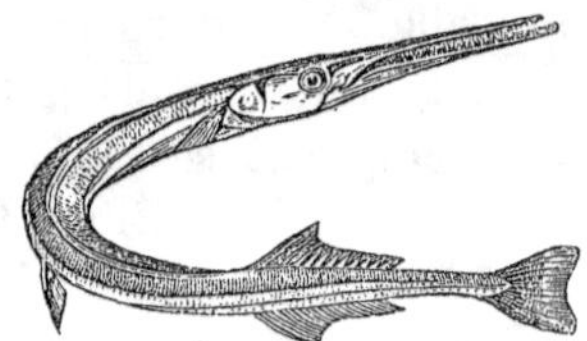

Fig. 215. — Gar-Fish.

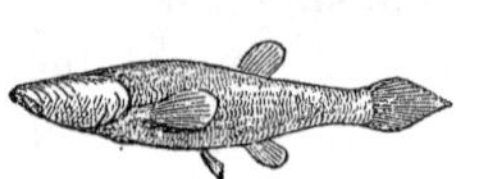

Fig. 216. — Blind Fish.

Fig. 217. — Horned Pout.

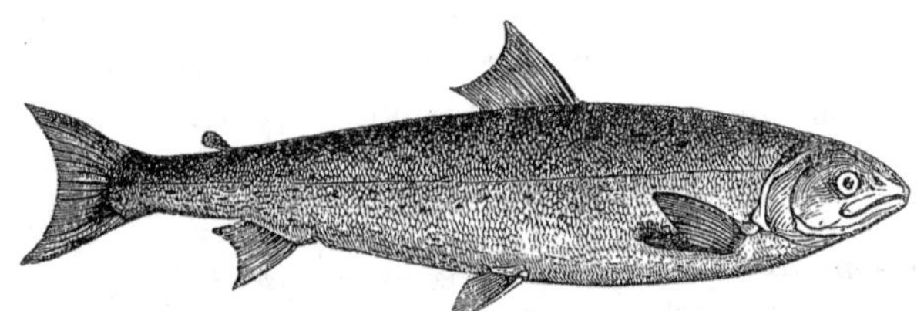

Fig. 218. — Salmon.

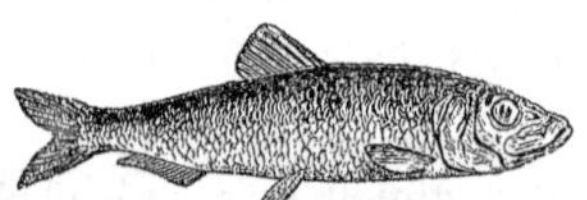

Fig. 219. — Herring.

Fig. 220. — Speckled Trout.

teeth over their whole inner surface, and a row of long, pointed teeth on their edges. It is found in the Northern lakes and in the Western and Southern rivers.

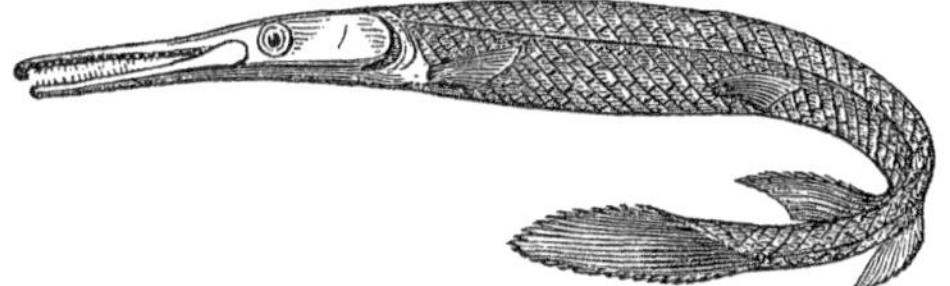

Fig. 221.—Gar-Pike.

The Cod inhabits the North Atlantic, and attains a weight of even a hundred pounds in some cases. It is taken in immense numbers on the Banks of Newfoundland, and when salted and dried is carried to all parts of the world. The Flounders are marine fishes which have the body flattened on the sides, and both eyes are on the same side of the head, and the side upon which the eyes are placed is always uppermost, and is dark colored, while the opposite side is white. They swim, therefore, on one side, and they keep close to the bottom. Flounders are from six inches to two feet long, and are caught in great numbers, even from the wharves. Halibuts are shaped like the Flounders, and in some cases weigh six hundred pounds. The Flounders and the Halibuts are the only backboned animals which have the right and left sides unlike. The Lump Fishes are those whose ventral fins are so joined as to form a sort of cup, by which they are able to attach themselves firmly to rocks or other objects. Pennant, the naturalist, says that he put one into a pail of water, and it adhered so tightly to the bottom that he lifted the whole pailful by taking hold of the fish by the tail. It lives in the North At-

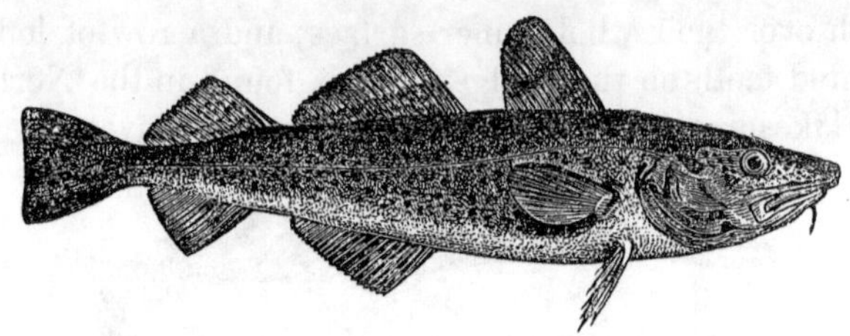

Fig. 222. — American Cod.

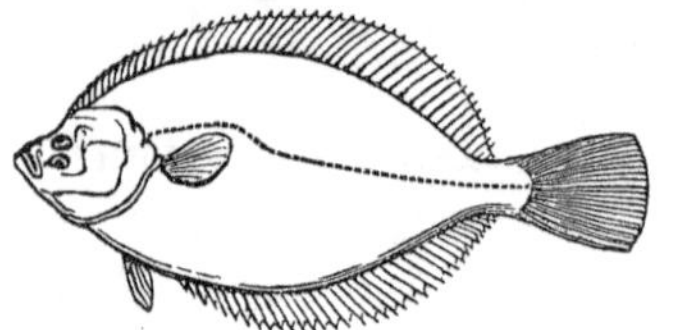

Fig. 223. — Flounder.

Fig. 224. — Burbot.

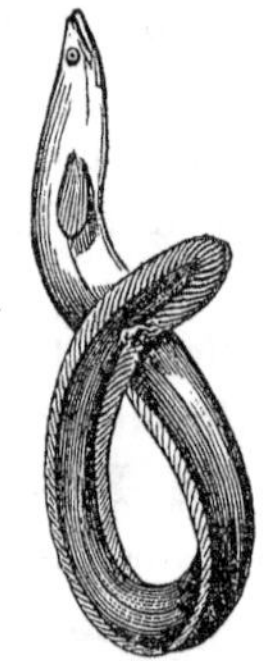

Fig. 225. — Eel.

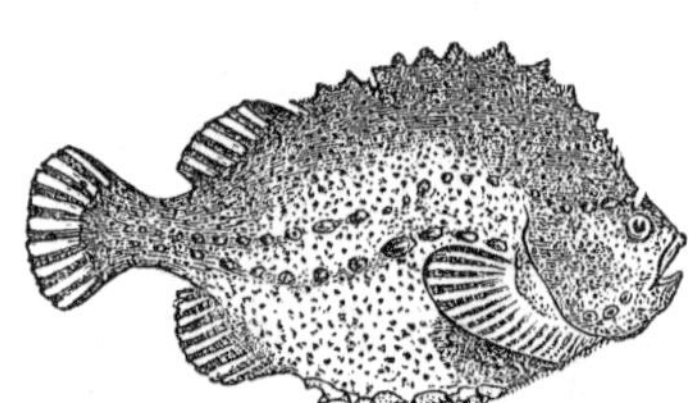

Fig. 226. — Lump Fish.

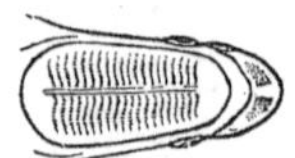

Fig. 227. — Top of Head of Remora.

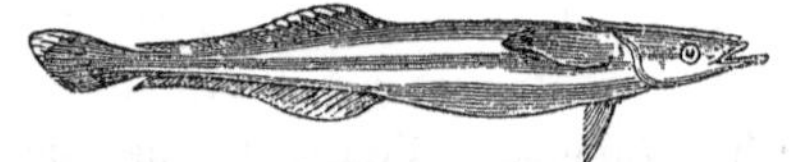

Fig. 228. — Remora.

lantic. The Remora has a flattened head, so constructed that it is able to attach itself by this part to other marine animals. It is a foot or more in length. Eels have a long, round body, which is covered with a thick, soft skin. They live in both fresh and salt waters, and keep near the bottom, often lying concealed in the mud.

TUFT-GILLED FISHES.

These fishes have their gills in tufts, and are known as Pipe-Fishes and Sea-Horses, on account of their singular forms. Pipe-Fishes have a very long and slender body covered with hard plates, and long snout with the mouth at the end. They live in the warm seas. After the eggs are laid, the male takes them in a sort of sack and carries them about with him till they are hatched. Sea-Horses have a short body, which is covered with spiny plates, a tail adapted for grasping small objects, and the head and neck have some resemblance to those of a horse. They are from three to six inches long, and live in the sea.

PUFFERS, TRUNK-FISHES, &c., OR PLECTOGNATHS.

Puffers have the body covered with spines, and can swell themselves like a balloon by swallowing air. The Common Puffer lives in the Atlantic Ocean, and is about a foot long.

The Sun-Fish, of the Atlantic, grows to the length of four feet, and weighs five hundred pounds.

The Trunk-Fish has the head and body covered with bony plates, so firmly attached to each other that they form a shield, and the mouth, tail, and fins are the only movable parts. Two or three kinds are found on the Atlantic coast of the United States.

Fig. 229. — Sea-Horse.

Fig. 230. — Pipe-Fish.

Fig. 231. — Puffer.

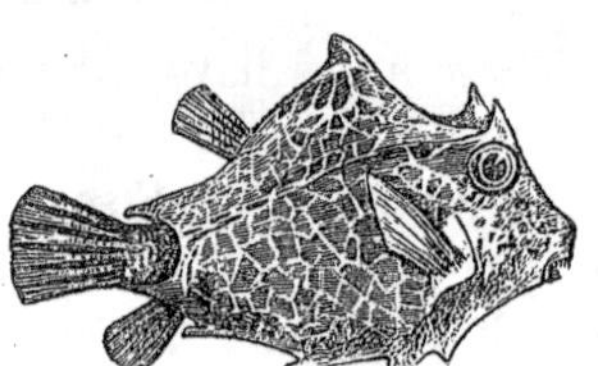

Fig. 232. — Trunk-Fish.

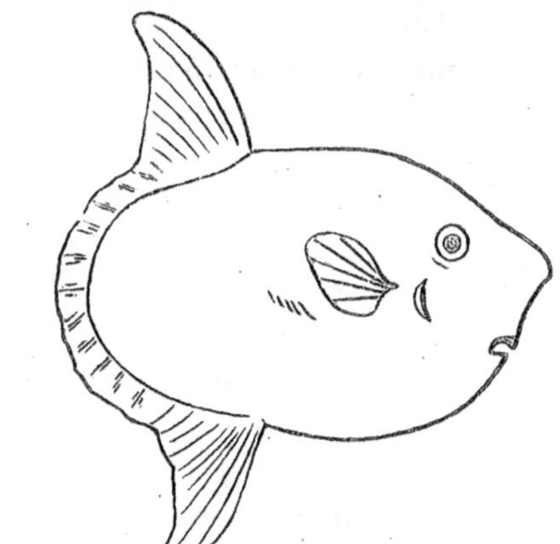

Fig. 233. — Sun-Fish.

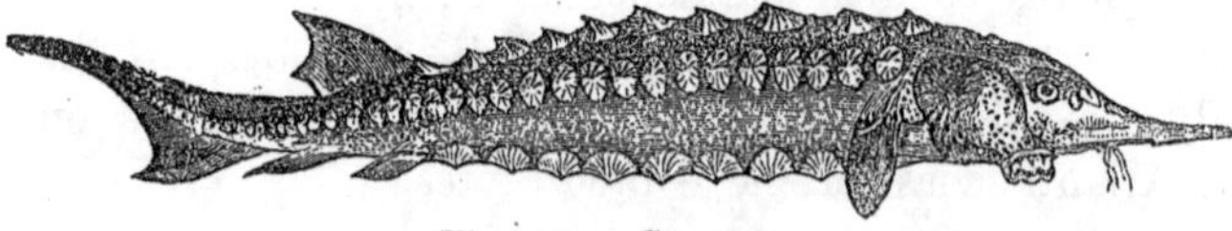

Fig. 234. — Sturgeon.

STURGEONS.

Sturgeons are fishes whose skeleton is a sort of cartilage, instead of being bony, as in those already described. They are also covered with bony plates placed in rows along the whole length of the body, and the mouth is under the snout, and can be much protruded. They inhabit lakes and the ocean, and ascend rivers. They are from three to ten feet long. See Figure 234.

SHARKS, OR SELACHIANS.

These are marine fishes which have the skeleton cartilaginous, and which, in many cases, are very large, and in most cases very ferocious. The different kinds vary from four to thirty feet in length, and their teeth are very numerous, sharp as lancets, and inflict the severest wounds. The smaller marine animals and even men fall a prey to them.

The Rays, or Skates, have the body broad and flat, and are from two to six feet or more in length, and as wide or wider than the length. Those called Vampires are sometimes sixteen feet wide, and weigh several tons! One kind of Ray, called Torpedo, gives violent electrical shocks when touched. See Figures 242 and 243.

SUCKERS, OR CYCLOSTOMES.

The true Suckers are the least perfect or lowest of all the fishes, and their tongue moves forwards and backwards like the piston in a pump, enabling them to produce a vacuum, and thus to fix themselves to other fishes. The Sea Lamprey, two or three feet long, the Hag or Myxine, six or eight inches long, and the Amphioxus, or Lancelot, are of this kind. The last two also live in the sea, and are seldom seen. The Lamprey ascends rivers, and piles up heaps of stones, among which it lays its eggs. See Figures 244–246.

Fig. 235. — Thresher Shark.

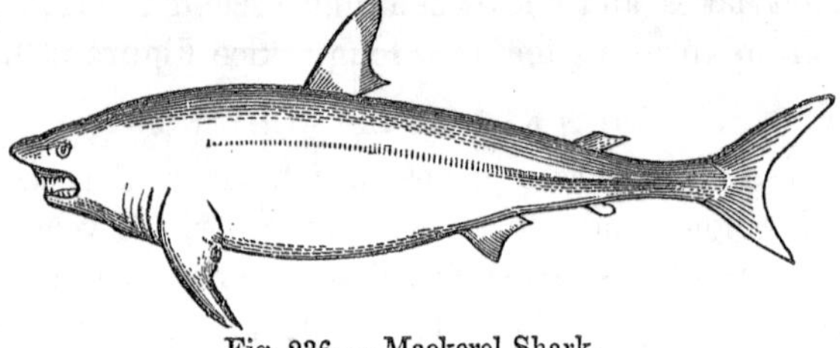

Fig. 236. — Mackerel Shark.

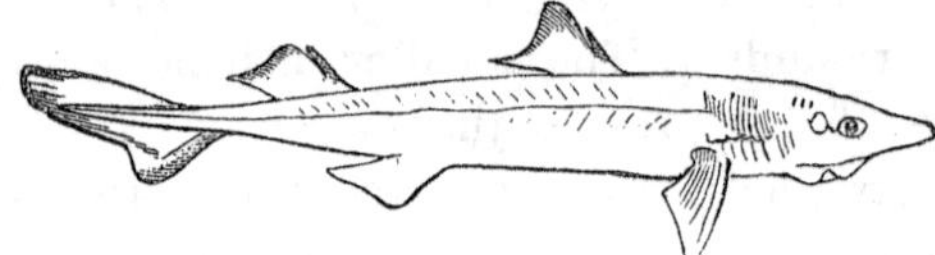

Fig. 237. — Dog-Fish Shark.

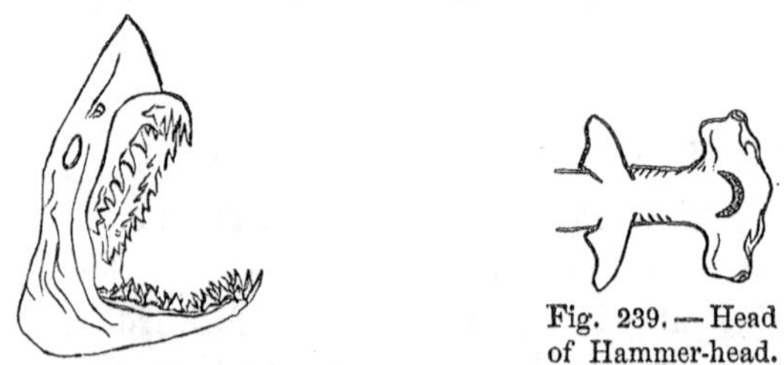

Fig. 238. — Head of Mackerel Shark.

Fig. 239. — Head of Hammer-head.

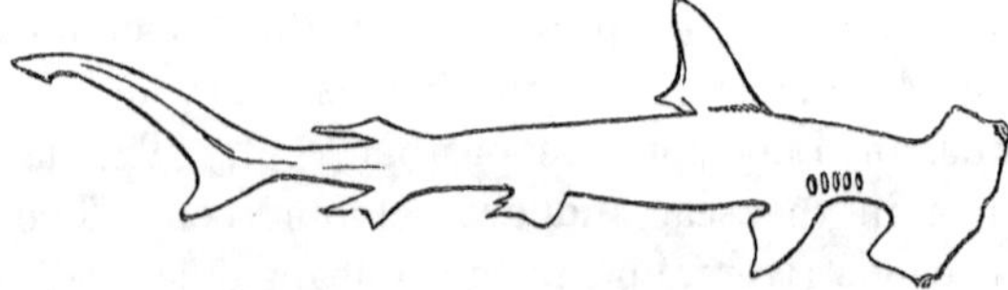

Fig. 240. — Hammer-head Shark.

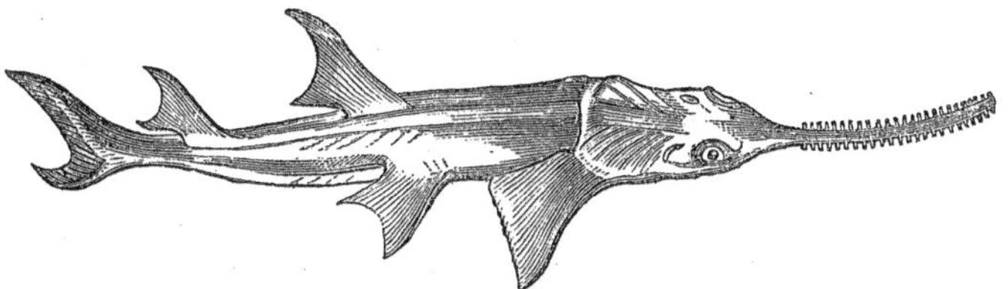

Fig. 241. — Saw-Fish. A Shark.

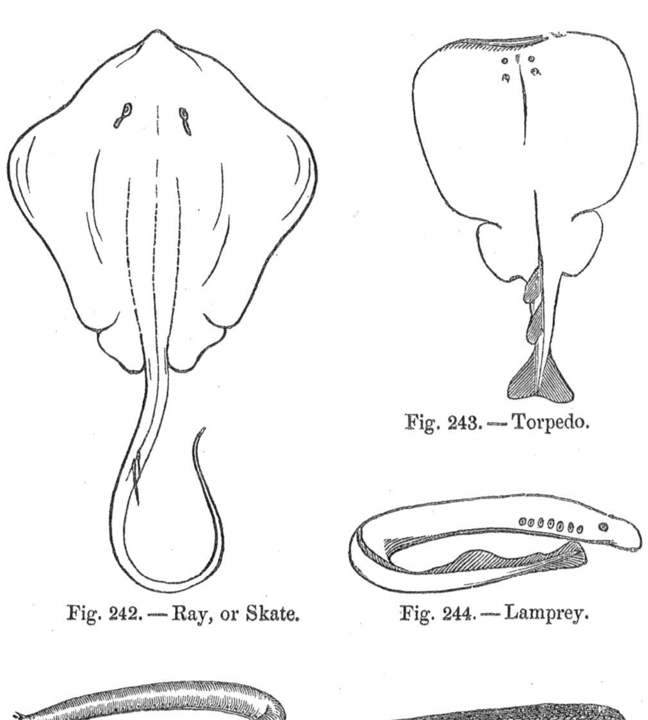

Fig. 243. — Torpedo.

Fig. 242. — Ray, or Skate.

Fig. 244. — Lamprey.

Fig. 245. — Hag or Myxine.

Fig. 246. — Amphioxus.

ARTICULATES, OR JOINTED ANIMALS.

THESE are animals which have no internal skeleton, nor backbone, but whose hard parts, when these exist, are upon the outside, and whose body is made up of a series of similar rings. Articulates include the Insects,—Bees, Butterflies, Flies, Beetles, Bugs, Grasshoppers, Darning-Needles, &c.,—Lobsters and Shrimps, and Worms.

INSECTS.

Insects are articulates which breathe by means of air-holes along the sides of the body; and these air-holes are the openings of air-tubes which branch throughout the body and carry air to every part. The term *Insect* comes from a Latin word which means *cut into*, and is given to these animals because they seem to be cut into, or notched. True Insects, Spiders, and Centipedes are of this kind.

True Insects have the body divided into three parts, —the head, middle body or thorax, and hind body or abdomen. Upon the head, and near to the eyes, are placed two jointed members, called antennæ, which it is supposed are connected with the sense of hearing, or of touch, or of both of these senses; to the middle body or thorax are attached the legs and wings; and the hind body contains the organs of digestion, and to this part belongs also an organ called the sting, or piercer. Insects either bite their food or suck it. Those which bite their food have an under and upper lip, between which are two pairs of jaws which move sidewise, and four to six little feelers, which they use to touch and examine the food. Those insects which suck their food have either a long tube, as Butterflies and Moths; a piercing sucker, as Mosquitoes; a softer one, used for lapping, as Flies; or a jointed one, which is doubled under the breast when not in use, as Bees. The eyes of insects appear to be only two in number, but each is composed of many single eyes — often thousands, and in some cases the astonishing number of twenty-five thousand — closely united. Many winged insects have also one, two, or three single eyes on the crown of the head. The legs are six in number, and are attached to the under side of the thorax; the wings are two or four, and vary greatly in form and thickness, in veinings, and in the manner of folding when at rest. The hind body is the largest portion, and the air-holes are found on this part. Insects are produced from eggs. A very few do not lay their eggs, but retain them in the body till hatched; others always lay their eggs where the young will find a plentiful supply of food. Most insects undergo great and wonderful changes in form and habits; so great, that the same insect, at different ages, might be taken for as many different animals. For example: a caterpillar, after feeding upon leaves until it is fully grown, casts off its skin, and appears as a much smaller, oval body, which neither moves about nor takes food. After remaining awhile in this state, there is motion within, the skin bursts open, and there comes forth a butterfly, or a moth, whose wings expand, and harden, and are soon able to bear it away in search of flowers, upon whose honey it feeds. In its first state it is called a *larva*, — a word which means a mask, — because its future form is masked, or concealed; in the second

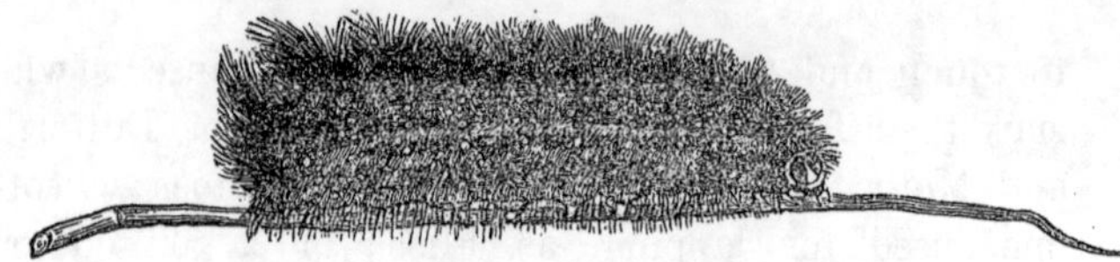

Fig. 247. — Larva.

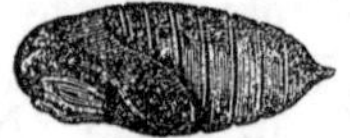

Fig. 248. — Pupa of Fig. 247.

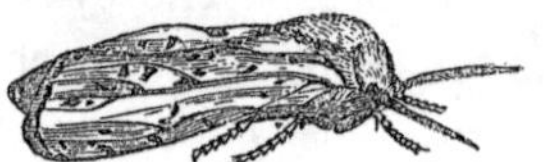

Fig. 249. — Imago of Figs. 247, 248.

Fig. 250. — Larva.

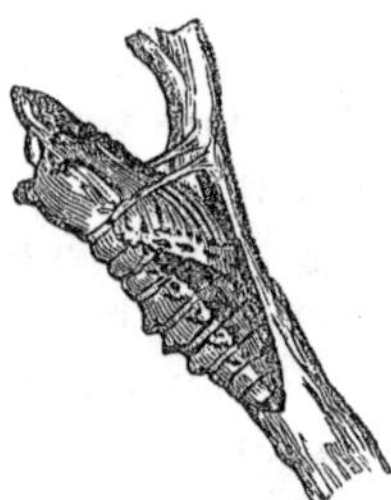

Fig. 251. — Pupa of Fig. 250

Fig. 252. — Imago of Figs. 250, 251.

state it is called a *pupa*,—a word meaning infant,—from a slight resemblance that some insects in this state bear to an infant clothed with bandages, according to a custom among the Romans; and it is also often called a *chrysalis*, from a Greek word which means gold, because some of the pupæ are adorned with golden spots; in the third state it is called a perfect insect, or *imago*, from a word which means image, because the image concealed in the skin of the pupa has come forth. These different states are plainly shown on page 127. Some caterpillars spin a silken covering, which is called a *cocoon*, from a word which means a shell; all the silk of the world comes from the cocoons of these little creatures. Insects which pass through the changes just described are said to undergo a complete transformation; but there are some insects which do not change their form so completely. Grasshoppers, for instance, are active during their whole lives, never passing through an inactive pupa state. When hatched from the egg they have legs, but no wings; later their wings begin to grow, and at length, having shed their skin several times, each time appearing with longer legs and more perfect wings, they reach their full growth, shed the skin for the last time, and appear as perfect, or adult grasshoppers. Such insects undergo only a partial transformation.

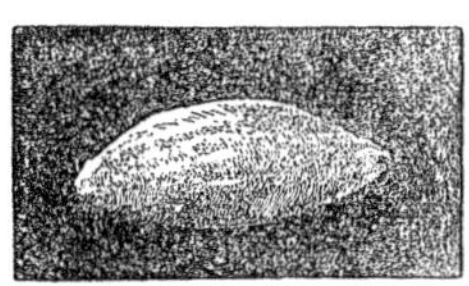

Fig. 253.—Cocoon.

Insects are the most numerous of all the classes of animals, there being several hundred thousand kinds. And the study of Insects is one of the most interesting and fascinating in which one can engage. The study of Insects is also very important, that we may know which are injurious to the farm, orchard, and garden, and to the granary and closets, and by knowing their habits be able to resist their attacks; and that we may know which are of use to man; for God has so made the Bee that it gives us delicious honey, and some of the Beetles, so that they are of use to the sick, some of the little Bark-Lice, as the Cochineal, so that they yield the richest dyes, and some of the Caterpillars, so that they may furnish all the world with silk.

BEES, WASPS, ICHNEUMONS, &c., OR HYMENOPTERS.

These insects have four wings which are more or less transparent, the hind pair being the smaller, and all with a few branching veins. They have two pairs of jaws,—the upper pair fitted for biting, and the lower pair with the lower lip adapted for collecting honey. The females have either a sting or a piercer. They surpass all other insects in the number and variety of their instincts. The word Hymenopter means *membrane-winged.*

BEES.

Bees have a hairy body, and their mouth is lengthened into a sort of proboscis, which is jointed and can be folded under the head, and the first joint of the hind legs is often very large, and fitted for collecting and carrying the pollen of flowers.

The Hive or Honey Bee is originally from Asia, but has now spread over Europe and America. It is seen almost everywhere in hives, and it is also quite common in a wild state, and often far from human dwellings. In a wild state, Bees of this kind have their home in hollow trees and in clefts of rocks. In every nest or hive there are three kinds, a female or queen, males or drones, and workers In a well-stocked hive there are two thousand males, fifty thousand workers, but only *one* queen. The workers are the smallest;

Fig. 254. — Queen.

Fig. 255. — Worker.

Fig. 256. — Drone.

Hive Bee.

they fly over the surrounding country and collect all the materials to form the structure called the comb; they build the cells and store them with honey; they feed and protect the young; they wait upon the queen; they do all the work of the hive. The males or drones have a thicker body, and no sting; they perform no labor, but are supported by the workers. The queen is much larger than the others, has a sting, and is the sole mistress of the hive. She lays all the eggs, and seldom goes out except to lead a swarm. The honey-comb is one of the most interesting of insect structures, and is arranged in the hive in the most regular manner. The cells are six-sided, and are built in just the shape to save all the room, and to be the strongest, and to contain the greatest amount of honey, and in the shape which requires the least amount of wax in their construction.

There are certain cells in which the queen lays her eggs, depositing one in each cell; and when the eggs are laid, the workers fill the cells with the pollen of flowers mixed with water and honey, — this is food for the larvæ. In about two days the eggs hatch into small white larvæ, and in five or six days these begin to spin a cocoon, and soon go into the pupa state. A queen comes forth from this state in sixteen days, workers in twenty days, and drones in twenty-four days. As only one queen can live in a hive, whenever a young queen is hatched she is carefully guarded from the old one by the workers, till it is settled whether the old queen will be wanted to lead forth a swarm. If a new swarm is not to go forth, the old queen is allowed to approach the young queen and royal cells, and destroy the brood, which she does by stinging them. If the old queen leaves with a swarm, the young queen immediately endeavors to destroy her sisters, but is prevented by a guard of workers, while there is a prospect of another swarming; if she departs with a swarm, another queen is set free, and so on till further swarming is impossible; then the young queen is allowed to kill all her sisters. If two queens hatch at the same time, they instantly engage in conflict, the other bees favoring the battle, and when one is killed, the survivor is recognized as queen. When a hive loses its queen, there is the greatest confusion; after several hours they become quiet, and if there are no eggs or larvæ in the cells from which a new queen may be hatched, they become discouraged, cease to labor, and the whole colony soon dies. If there be eggs or larvæ in the cells, the bees select one, — the larva of a worker, — and destroying the cells adjoining, so as to make a royal cell, they supply the grub with the sort of food prepared for queens, and in this way soon raise another queen.

The Humble Bees are larger than the Hive Bees, and their bodies are very hairy. There are more than forty kinds in North America. They build nests in the ground, or under stones, or in deserted mouse-nests, and their cells are large and egg-shaped. Sometimes there are four hundred bees in a community, the descendants of one female bee which lived through the winter and founded the colony in the spring. The Carpenter Bees are also large, and they cut tubular holes in posts and stumps, and lay their eggs there, arranging them in layers of the pollen of flowers. The Mason Bees make their nests of sand, in crevices.

WASPS.

Wasps usually live in colonies composed of males, females, and workers. Unlike Bees, they prey upon other insects. They build nests under ground, or in holes, or attach them to bushes, trees, fences, or buildings. The nest is usually made of a substance which they gnaw from wood, and which, by the action of their

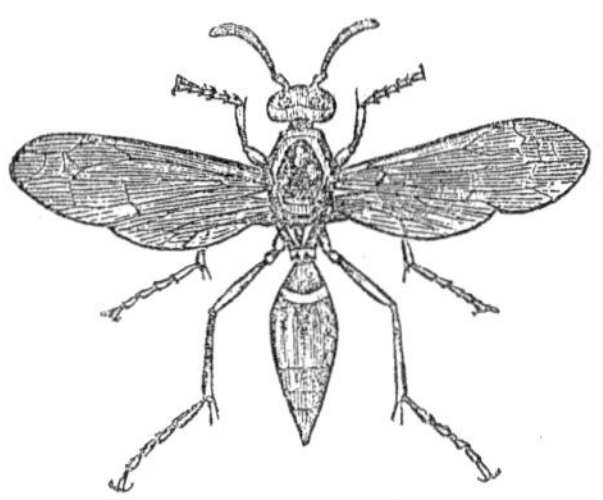

Fig. 257. — Wasp.

jaws, they reduce to a pulp, which hardens into a sort of paper. The Wasps were the first paper-makers, and they were the first to show that paper can be made of wood. The combs lie horizontally in the nest, and are made of the same paper-like material as the nest, and each is attached to the one below it by a sort of pillar. The cells contain no honey, but are built for places in which to rear the young. The colony is dissolved on the approach of winter, the males die, and the females seek a sheltered winter retreat. Each female that survives the cold founds a new colony in the spring, builds a few cells, lays her eggs, from which are hatched only workers, which assist the parent, and at length, in autumn, three generations have been produced, the last composed of males and females, and the nest has grown from a few cells to one containing thousands. The Hornet is one of the largest of the wasps, and was brought to this country from Europe.

Some kinds of wasps build an open nest of a few cells, and attach it to some object by a short stem. Other kinds build their nests of mud, and store them with insects for the food of the larvæ; these are the Mud-Wasps. They have the hind body joined to the thorax by a long stem or pedicel, and their color is shining blue, or black, or black and orange, or brown and red. One of the black and orange Mud-Wasps has just built two beautiful mud-cells in the corner of my room. She worked very industriously and rapidly, building a cell in a few hours. Flying in at the open window, with a ball of mud in her mouth, she moved quickly around the room, then flew up to the spot where she was building, and, depositing her mud, shaped it with her jaws with all the care and regularity of a perfect mason. The day after she finished the first cell, she filled it with spiders and sealed it over with mud. On opening it to examine the insects stored within, quite a large hole was accidentally made; this she very soon discovered, and began to repair it, and in about five minutes she had completely closed it. The second cell was soon finished, and sealed like the first. Here is a sketch of them, as they appeared before the second was filled with spiders and closed.

Fig. 258.—Mud-Wasp's nest.

One very large wasp, the *Stizus*, an inch and a half long, and of a black and yellow color, makes its nest in the ground, and stores it with insects. One of these, now in the Zoölogical Collection of Vassar College, was caught by S. M. Buckingham, Esq., in his garden, while carrying into its hole the Dog-day Cicada or Harvest-Fly, Figure 317.

ANTS.

Ants live together in colonies, which are often very large, and made up of males, females, and workers. The workers have no wings, but the males and females have wings, and the females have the power of throwing them off. Some kinds of ants make their nests in the ground; others raise large ant-hills; and others live in stumps and trunks of trees. The workers have the care of the nest and of rearing the young, they go abroad in search of food, communicate with and assist each other, feed the larvæ, and take them into the sunshine in fair weather, and back again on the approach of a storm, or at night, and watch over them earnestly and faithfully. Ants are fond of sweet things, and make pets of Aphides, or Plant-lice,—little insects which live upon the juices of plants, and yield a honey-like fluid. Some kinds of ants collect large numbers of aphides and keep them that they may eat the sweets which they produce. There is generally but one kind of ant in each nest, but in some cases the workers procure help by visiting the hills of other species, and forcibly taking the larvæ and pupæ, and bringing them back, where they are tended and reared by workers of the same kind which have before been stolen in the same way. Ants are very warlike, and engage in pitched battles, after which the ground is strewn with the dead.

ICHNEUMONS.

These insects have a long, hard, slender body, long antennæ, and the ovipositor is usually long; the latter is sometimes two or three times the length of the body. They lay their eggs in the eggs, larvæ, and pupæ of other insects, and thus destroy great numbers of them. Sometimes the eggs are laid upon the outside, but usually inside. When laid on the outside of the pupæ, the ichneumon, as soon as hatched, eats its way into its victim; when laid inside, the young ichneumon

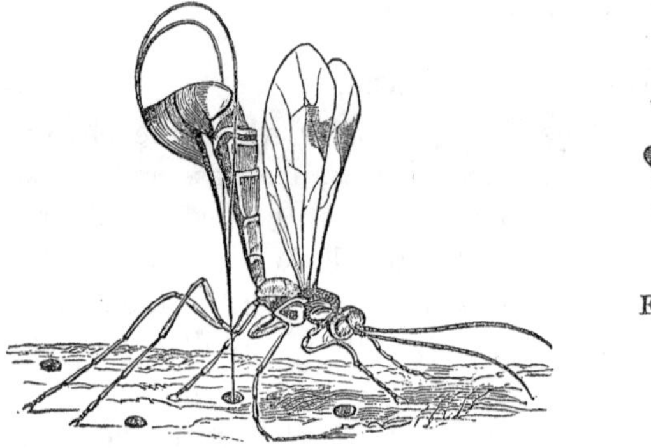

Fig. 259.—Ichneumon laying her eggs in holes bored by the Boring Saw-Fly, Figure 262.

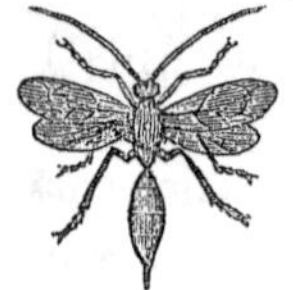

Fig. 260.—Ichneumon.

feeds upon the substance of the body, but attacks no vital part, so that the insect does not die till the ichneumon is ready to enter upon the pupa state.

GALL-FLIES.

These are very small insects, and the females have a long, slender ovipositor, with which they insert their eggs into leaves and other parts of plants. These punctures cause outgrowths called galls, which vary in size, form, and solidity, according to the nature or part of the plant that is wounded, and according to the kind of gall-fly that makes the wound. Some are shaped like an apple, as the gall of the oak; some like a bunch of currants; some are almost as hard as iron; and some are juicy and pulpy, like fruit. At length the eggs hatch, and the larvæ feed upon the vegetable matter which surrounds them. Some galls have only one tenant, others contain many, and usually these insects undergo all their changes within the galls, and, gnawing through the shell, fly away; but some kinds gnaw through at the end of their larval life, and enter the ground to go into the pupa state. Some produce no galls themselves, but live in galls made by others, and are called Guest Gall-Flies. The nut-galls used in making ink, in coloring, and in medicine, are caused by an insect which punctures a species of oak common in Western Asia. The Rose-bush Gall-Fly punctures the stems of rose-bushes, producing excrescences, or woody galls, upon them. One of the largest species is the Willow Gall-Fly; its galls are found at the ends of the twigs of the basket-willow.

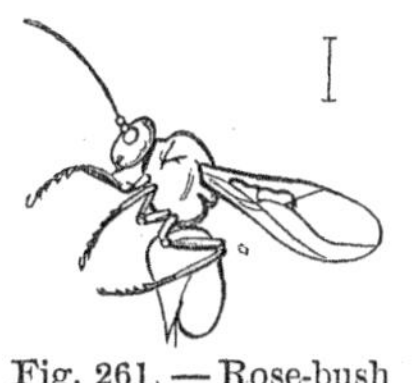

Fig. 261. — Rose-bush Gall-Fly.

BORING SAW-FLIES.

The Boring Saw-Flies have a long body, and the hind body is blunt, and ends in a horny point. Extending from beneath the hind body is a long, saw-

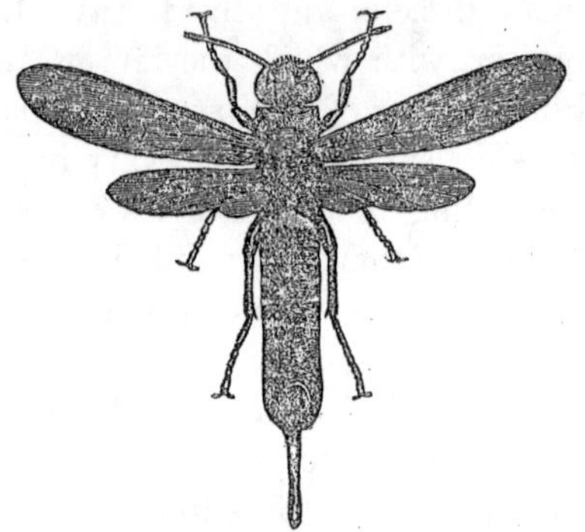

Fig. 262.—Boring Saw-Fly, or Pigeon Tremex.

like, and powerful borer, which is used to make holes in trees, in which to lay their eggs. In the larva state, they live in the trunks of trees.

TRUE SAW-FLIES.

The females of the true Saw-Flies have an ovipositor composed of two saws, enclosed between two horny pieces, which serve as a sheath. These saws are projected and withdrawn when the insect is cutting a place for her eggs; but not together, for while one is pushed forward, the other is withdrawn. When the hole is cut deep enough, the egg is deposited within. Saw-Flies

Fig. 263.—Fir-tree Saw-Fly, enlarged.

are sluggish, and fly only in the warmest days. The larvæ are found together in large numbers on the leaves of the birch and alder. When disturbed, they take very curious attitudes, appearing to stand upon the head, or they curl into the form of an S, or coil themselves up with the head in the centre, and look somewhat like a snail-shell.

BUTTERFLIES AND MOTHS, OR LEPIDOPTERS.

The word Lepidopter means *scaly-winged*, and is given to these insects because their wings are covered on both sides with minute scales. These are removed by the slightest touch, and to the naked eye look like a mealy powder; but when seen under a microscope, they are found to be little scales attached to the skin by a short stem. The tongue is long, and adapted for suction; when not in use it is rolled up like a watch-spring beneath the head, and partly concealed on each side by a little feeler. They have six legs, the first pair being short, and, in some cases, folded under the breast; and the feet end in a pair of claws. The young of Butterflies and Moths are called caterpillars, and these have from ten to sixteen legs. Most caterpillars feed upon the leaves of plants. Some eat buds, blossoms, seeds, and roots, and others eat the solid wood. Some eat woollens, others furs, others meat, lard, wax, and flour. Some kinds herd together in great numbers, and build nests in which they live, or to which they retire for shelter; others live in solitude, either in the light and air, or sheltered in leaves folded over them, or in silken sheaths which they make; and some conceal themselves in the ground, coming forth only to eat. In the middle of the lower lip there is a little tube, from which the caterpillar spins silken threads. Two long slender bags within the body, ending in the spinning tube, contain the material from which the silk is made; this is a sticky fluid, which

Fig. 264.—Turnus Butterfly.

hardens into a thread as soon as it comes to the air. Some caterpillars spin but little silk, others produce it in abundance. Caterpillars change their skins about four times in coming to their full growth as caterpillars; and when about to change into the pupa or chrysalis state, they cease eating, and many of them spin around their body a silken covering called a cocoon, others suspend themselves by silken threads, without making a cocoon, and others enter the ground. When the caterpillar is prepared for the change, it bursts the skin on the back, draws out the forward part of its body, and works the skin backward until it throws it off; and now it is a chrysalis, shorter than the caterpillar, and at first sight it appears destitute of head and limbs; but on looking more carefully we perceive traces of head, tongue, antennæ, wings, and legs. Some chrysalides are angular, but most of them are smooth, rounded at one end, and tapering at the other; they remain motionless, or only move the hind part of the body when touched. At length the enclosed insect is ready to come forth, and by many movements it bursts the skin of the back, and the Butterfly or Moth appears. At first it is soft, weak, and moist, with small and shrivelled wings; but soon the moisture passes off, the limbs become firm, the wings expand, and the perfect and beautiful insect goes forth to feed upon water and the honey of flowers. Butterflies and Moths do not increase in size; they are full grown when they emerge from the pupa skin; and after having laid their eggs, they soon die. Butterflies fly in the daytime, have their wings erect when at rest, and their antennæ are thread-like, with a little knob at the end, and their larvæ do not spin cocoons. Moths fly mainly at night, have their wings when at rest more or less sloping like a roof, and their antennæ are variously formed, but never knobbed at the end.

PAPILIO BUTTERFLIES.

The Turnus Butterfly is one of the largest in North America. Its color is a beautiful yellow, with black markings, and the hind wings are tailed. The caterpillar feeds upon the leaves of the apple and wild-cherry trees, folding them up so as to make a case for itself. When fully grown, it is about two inches long, green above, with rows of blue dots, and yellow and black marks, and its head and legs are pink. It becomes a chrysalis early in August, and comes out a butterfly the next summer. See Figure 264.

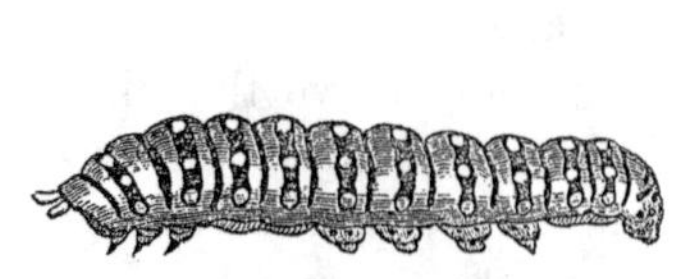
Fig. 265. — Asterias Butterfly,— Larva.

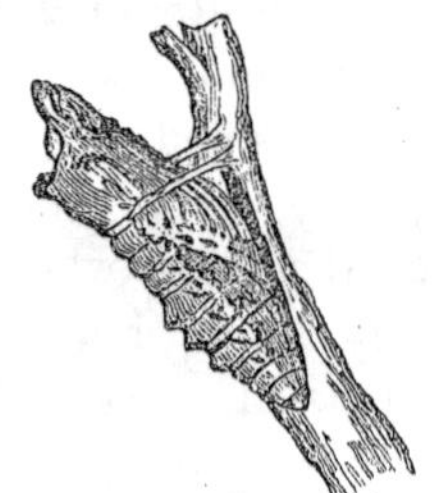
Fig. 266. — Pupa of Fig. 265.

Fig. 267. — Asterias Butterfly.

The Asterias Butterfly is black, with two rows of

yellow dots on the back, and two rows of yellow spots across the wings; the hind wings are tailed, and have seven blue spots between the two rows of yellow ones, and an eye-like spot of an orange color with a black centre. The female is much larger, and has fewer yellow spots on the upper surface. The caterpillar lives upon such plants as the carrot, parsnip, celery, and anise. It is green, with a band made up of yellow and black spots on each ring. When touched, it thrusts out from the head a pair of soft, orange-colored horns, which have a disagreeable odor. In July it reaches its full growth as a caterpillar; then it seeks a sheltered spot on the side of a building or fence, spins a tuft of silk, fixes its hind feet in it, then makes a loop of silk, and, passing its body through it, rests upon it as a support; soon it casts its caterpillar skin and becomes a pupa or chrysalis, Figure 266. In about a fortnight the chrysalis bursts open, and the perfect butterfly, Figure 267, appears.

WHITE, AND YELLOW BUTTERFLIES.

The Philodice, or Yellow Butterfly, expands about two inches, and is common in fields and by roadsides throughout the summer. The White Butterfly, or Pieris, is of about the same size, and is also common.

NYMPHALIS BUTTERFLIES.

These butterflies are remarkable for their beautiful colors. The Misippus Butterfly has the wings tawny yellow, veined with black, and a black border spotted with white, and the fore wings have near their tips a black patch spotted with white, and on the hind wings there is a curved black band. The caterpillar lives on the poplar and willow; its color is pale brown, marked with white on the sides, and on the second ring are

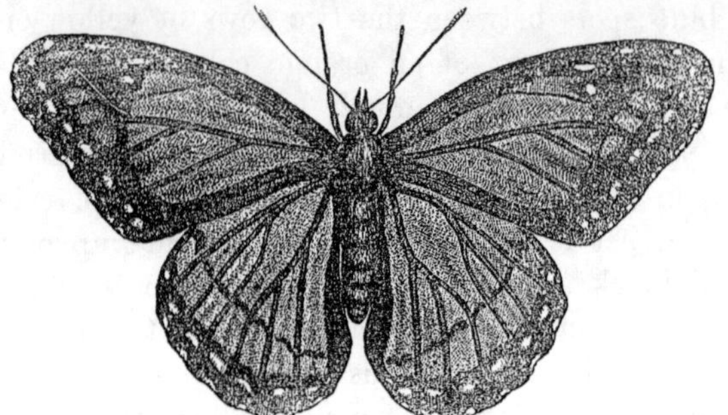

Fig. 268. — Misippus Butterfly.

two blackish horns. The butterfly is seen in June and September.

Other North American Butterflies are the Arthemis, Archippus, Idalia, Aphrodite, Bellona, Myrina, Phæton, Tharos, Thistle, Hunter's, Atlanta, Lavinia, Antiopa, White-J, Milbert's, Semicolon, Comma, &c., each of which is beautiful, and has an interesting story, but which must be omitted here for want of room.

SATYRUS BUTTERFLIES.

These have their wings broad and rounded, and those called Hipparchians have the wings of a delicate brown color, with beautiful eye-spots. They are very common in groves and about bushes late in the summer. Closely related to these is the Mountain Butterfly, which is found only on Mount Washington, in New Hampshire.

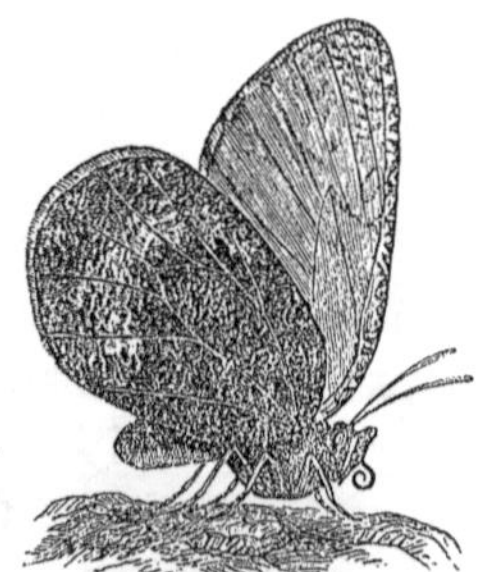

Fig. 269. — Mountain Butterfly.

SKIPPERS.

Skippers are butterflies which have a short body, large head, and large eyes; and the antennæ have the knob at the end either curved like a hook or ending in a little point bent to one side. They are called Skippers because they fly with a jerking motion. They are generally of a rich brown color, marked with spots of yellow, and expand from an inch and a half to two inches and a half.

Fig. 270. — Skipper.

The Tityrus Skipper is one of the largest and most beautiful species. Its wings are brown, and the forward ones have a yellow band across the middle and yellow spots near the tips, and the hind wings have a broad, silver-colored band across the middle of the under side. It is found about clover and other flowers in June and July. The females lay their eggs on the leaves of the locust-trees. The caterpillar, when full grown, is about two inches long, of a pale green color, with cross streaks of darker green; the head and neck are red, with a yellow spot on each side of the mouth.

AZURE AND COPPER BUTTERFLIES.

These are small, expanding only an inch. The Azure Butterflies are blue or brown; and the Copper Butterflies red, spotted with black.

HAWK-MOTHS, OR SPHINGES.

These moths are large, and have the antennæ thickest in the middle and usually hooked at the tip, and the wings long and narrow. During the morning and evening twilight, they may be seen flying from flower to flower with great swiftness, and are easily mistaken

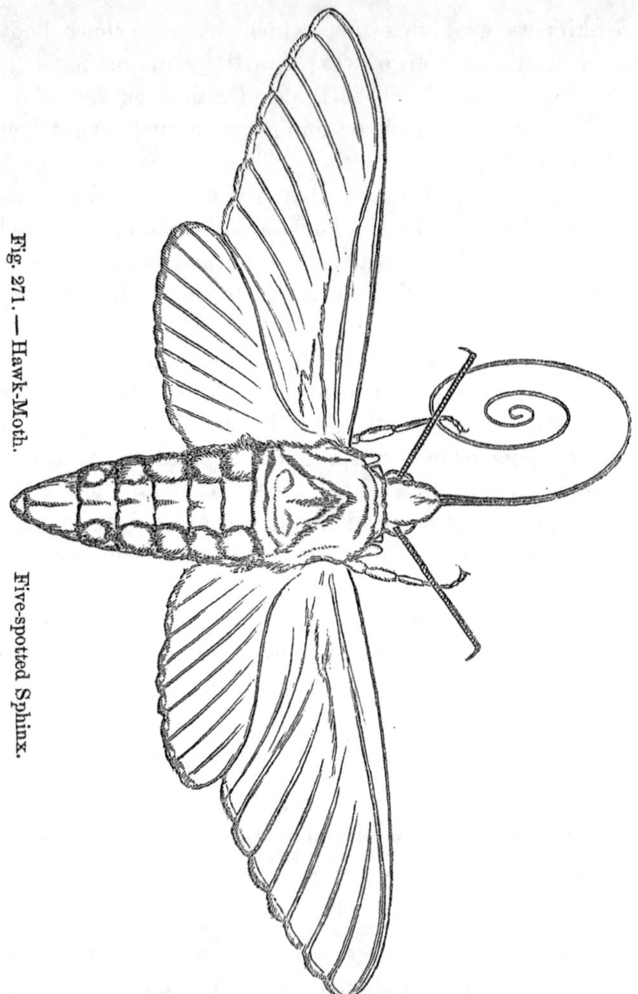

Fig. 271.—Hawk-Moth. Five-spotted Sphinx.

for humming-birds. A few kinds fly by day and in

bright sunshine. The caterpillars are very large, and are remarkable for their curious attitudes, which reminded Linnæus of the Sphinx, a sculptured monster of the Egyptians.

The Five-spotted Sphinx expands about five inches, and is of a mixed grayish and blackish color, and on each side of the body there are five orange-colored spots surrounded by black. Its tongue, when fully unrolled, is five or six inches long, but when not in use is coiled up nearly out of sight. The caterpillar is known as the potato-worm, and is green, with oblique whitish stripes on the sides, and a thorn-like projection on the tail. It attains its full length, three inches or more, in August, and then buries itself in

Fig. 272. — Larva of Five-spotted Sphinx.

the ground. Here, in a few days, it throws off its skin and becomes a chrysalis, of a bright brown color,

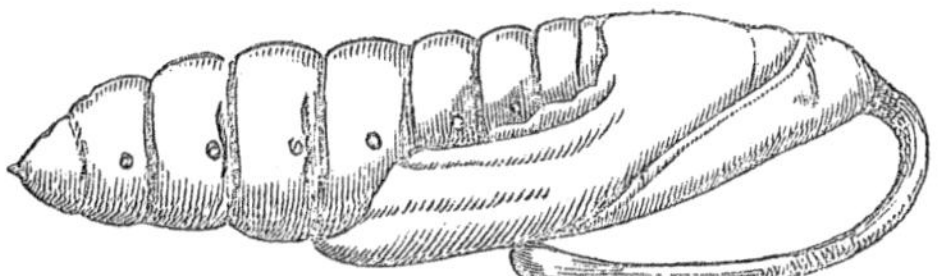

Fig. 273. — Chrysalis of Five-spotted Sphinx.

with a long tongue-case bent over from the head, its end touching the breast, and somewhat resembling the handle of a pitcher. It remains in the ground all winter, and in the following summer the large moth crawls out of it, comes to the surface, mounts a plant, and waits till the approach of evening, when it flies away in search of food.

CLEAR-WINGED SPHINGES, OR SESIAS.

These are known by their transparent wings and

Fig. 274. — Clear-winged Sesia.

broad tails. They are seen in the daytime hovering over flowers, like humming-birds. They are very beautiful.

PEACH-TREE BORERS.

The Peach-tree Borer, in its winged form, resembles a wasp. The general color is steel-blue, with yellow markings, and the male has all the wings transparent; but the female has the fore wings blue and opaque. The eggs are laid upon the trunk of the tree, near the roots. When hatched, the larvæ bore into and devour the inner bark and sap-wood. When about a year old they make their cocoons, become chrysalides, and come forth as moths from June to October.

Fig. 275. — Peach-tree Borer.

SILK-WORM MOTHS.

These moths have the head small, antennæ generally feathered or toothed, tongue short, thorax woolly, and the fore legs hairy. Most of the caterpillars spin cocoons. Some of these moths are small, and others are the largest of the Lepidopters.

One of the most elegant kinds is the Beautiful Deïopeia. Its fore wings are yellow, crossed by white

Fig. 276. — Beautiful Deïopeia.

bands, on each of which is a row of black dots, and the hind wings are scarlet with an irregular black border.

The Salt-marsh Moth expands about two inches; the fore wings are white, hind wings and hind body yellow,

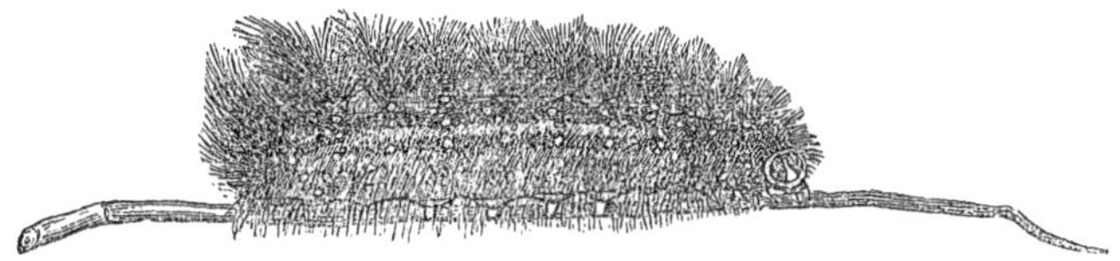

Fig. 277. — Salt-marsh Moth, — Larva.

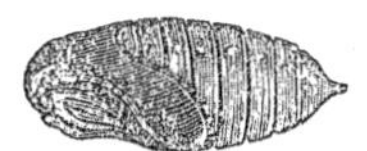

Fig. 278. — Salt-marsh Moth, — Pupa.

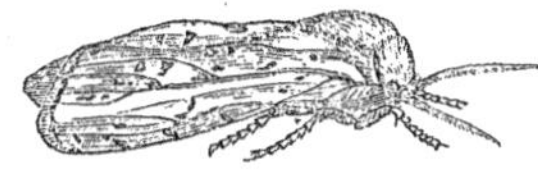

Fig. 279. — Salt-marsh Moth.

and the wings spotted with black, and the hind body has a row of black spots above, a row below, and two

rows on each side. The female has all the wings white, or all light gray, with the black spots.

The common Silk-Worm is celebrated as the insect which produces the greater part of all the silk used in the world. It is the larva or caterpillar of a moth, —*Bombyx mori*,—which expands about two inches, and which is of a light color, with two or three obscure streaks, and a spot on the upper wings. It feeds upon the leaves of the mulberry-tree, and spins a cocoon about an inch and a half long, of a yellow color, and which contains about one thousand feet of silk. This silk-worm is a native of China, but is now raised extensively in Europe, and, to some extent, in this country. The larvæ of several other moths, most of them of large size, are now raised, not only in Asia, but also in Europe and in the United States, for the purpose of producing silk.

The Cecropia Moth, the Promethea Moth, the Luna Moth, and the Polyphemus Moth are all large and magnificent species,—the largest in North America. They have the antennæ broadly feathered on both sides, and beautiful eye-like spots on the wings. All but the Promethea expand five or six inches, and the latter expands about four inches. They appear in June. The Cecropia is dusky brown, and near the middle of each wing is a dull red spot with a white centre and a narrow black edging, and beyond the spot a dull red band bordered on the inside with white, and near the tips of the fore wings is an eye-

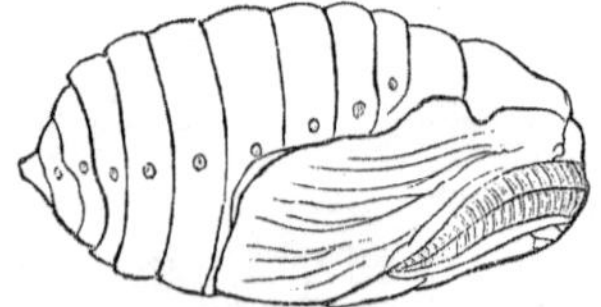

Fig. 280.—Chrysalis of Cecropia, cocoon removed.

like black spot. The caterpillar is light green, with red and yellow warts covered with short bristles. The cocoon is very large, three inches long, and fastened to the side of a twig; the outer coat looks like strong brown paper, and inside of this is loose strong silk surrounding an inner cocoon, which contains the chrysalis.

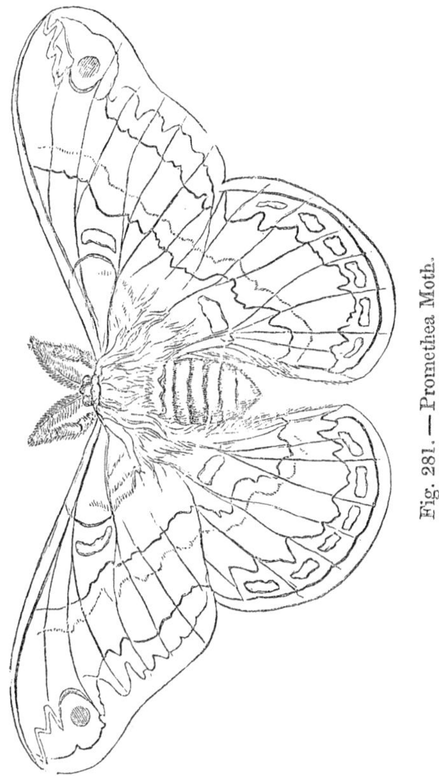

Fig. 281. — Promethea Moth.

The Promethea is brown with a wavy whitish line near the middle, and with a wide clay-colored border, marked by a wavy reddish line, and near the tips of the fore wings there is an eye-like spot. The caterpillar feeds upon the sassafras-tree. Before making its cocoon, it fastens to the twig, with silken threads, the leaf that is to cover its cocoon, so that it shall not fall in autumn; then it spins its cocoon on the leaf, bending over the edges to cover it.

The Luna, or "Pale Empress of the Night," is of a delicate light green color, and the hind wings are prolonged into a tail, and each wing has an eye-spot, which is transparent in the centre and surrounded by rings of white, red, yellow, and black. The caterpillar lives on the walnut and hickory, and is bluish green, with a yellow stripe on each side, and yellow stripes across the body. It draws together two or three leaves and spins its cocoon inside of them. The cocoon falls with the leaves in autumn, and the next June the beautiful Luna appears.

The Polyphemus Moth is reddish yellow, with a transparent eye-spot, divided by a slender line and encircled by yellow and black, on each wing, and on the hind wings adjoining the eye-spot is a large blue spot shading into black.

The American Tent-Caterpillar Moth expands an

Fig. 282. — Tent-Caterpillar Moth.

Fig. 283. — Cocoon of Fig. 282.

inch and a half, and is reddish brown, the fore wings crossed by two oblique whitish lines. The caterpillars abound in neglected orchards and upon wild-cherry trees. The eggs from which they hatch are placed in a cluster on the smaller branches, and are covered with a water-proof varnish. They hatch about the time the leaves unfold. The little caterpillars immediately form a small tent between the forks of the branches. As they grow, they enlarge the tent, surrounding it with new layers. They feed at stated times, and return to their tents when they have finished eating. In crawling from one twig to another they spin a silken thread, which serves to guide them on their way back. They rest in their webs at noon and in stormy weather. When full grown, which is about the middle of June, they leave the trees, separate, wander about for a time, and at length, in some sheltered place, spin their cocoons, which are oval and loosely woven, and the meshes are filled with a thin paste, which, on drying, becomes a yellow powder. They remain chrysalides about fifteen days.

GEOMETERS, OR SPAN-WORMS.

The Geometers are moths whose caterpillars seem to measure the surfaces over which they pass. They are obliged to move in this way, because they usually have only ten legs, six true legs on the fore part of the body, and four prop legs at the hind extremity. Geometers live upon trees, and let themselves down to to the ground by a silken thread which they spin from the mouth while descending. When not eating, many of them stand on the hind legs, with the body extended, and in this attitude may easily be mistaken for a twig.

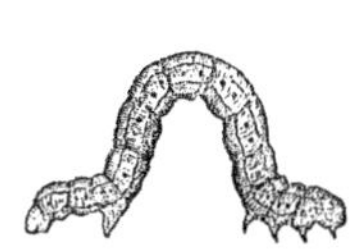
Fig. 284. — Geometer, or Span-Worm.

Often when disturbed, they let themselves down, hang till the danger is past, and then climb up by the same thread.

The Canker-worm Moth expands about an inch and a quarter, and the wings are large, thin, and silky. The females have no wings. The larvæ, called Canker-worms, the most destructive of insects, make their appearance about the time the leaves of the apple-tree begin to start from the bud. They hatch from clusters of eggs which have been placed upon the fruit and shade trees at various times in and since the autumn before. They immediately commence to eat. They first pierce the leaves with small holes, but as they grow they enlarge these holes, and by and by little more is left than the midrib and veins. When not eating, they lie stretched at full length beneath the leaves. When about four weeks old they reach their full size, about an inch long. They now quit eating, descend to the ground, and, entering to the depth of a few inches, each makes a little cavity, and soon passes into the chrysalis state. Here they remain till after the first frosts of autumn, when they begin to come forth in the moth state, and continue to do so, whenever the weather is mild enough, throughout the remainder of the autumn and the winter. They rise in the greatest numbers, however, in the spring. They come out of the ground mainly in the night. The females crawl up the nearest trees, where they are joined by the males, and soon begin to lay their eggs, which they place in rows, forming separate clusters of sixty to a hundred or more, each cluster being the product of a single female.

LEAF-ROLLERS.

The Leaf-Rollers are moths which, in the caterpillar state, roll up the edges of leaves, fastening them with threads of silk, and leaving the ends of the roll open. The moths are small, with the fore wings prettily banded, and sometimes adorned with golden spots.

Fig. 285.—Leaf-Roller.

TINEANS.

These moths, in the larva state, gnaw winding passages in the substances upon which they feed. They devour some of the fragments, and fasten together others with silken threads, thus making a covering for their tender bodies. They are the smallest of the Lepidopters, and are generally very beautiful. They enter through the cracks into closets, drawers, and chests, they get under the edges of carpets, and into the folds of curtains and garments, and here deposit their eggs. In about fifteen days the eggs hatch, and the larvæ immediately begin to gnaw whatever is within reach, covering themselves with the fragments, and shaping them into hollow rolls, and lining them with silk. They generally live in these rolls through the summer, but become torpid in autumn, change to chrysalides in spring, and in twenty days come forth winged moths.

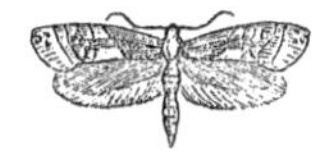

Fig. 286.—Tinean.

TWO-WINGED INSECTS, OR DIPTERS.

Flies, Mosquitoes, the Hessian Fly, Bee-Flies, Horse-Flies, and all their numerous relatives, have only two wings, the place of the hind wings being occupied by two small knobbed threads, called balancers. Mosquitoes have a long bill composed of bristles sharper than the sharpest needles, with which they pierce the flesh of men and animals, and secure the blood upon which they so much delight to feed. The female lays her

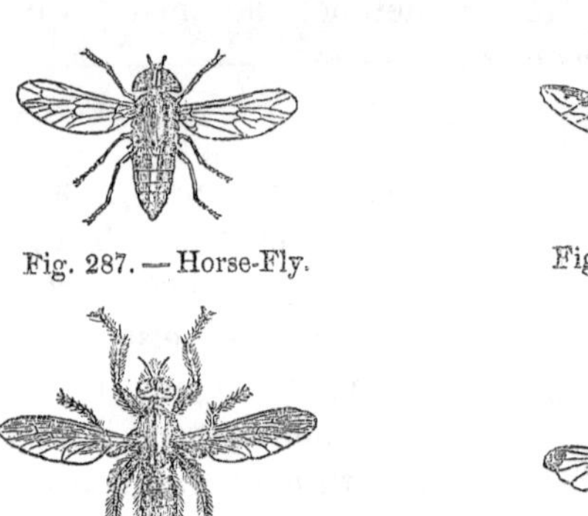

Fig. 287. — Horse-Fly.

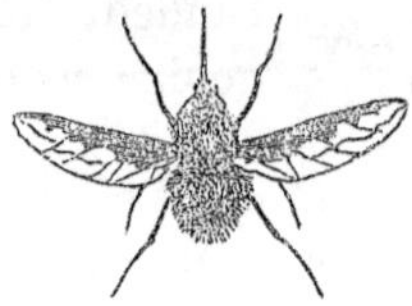

Fig. 288. — Bee-Fly.

Fig. 289. — Asilus-Fly.

Fig. 290. — Bot-Fly.

eggs on the surface of the water, and the larvæ may be seen in great numbers, throughout the summer, in all stagnant pools. They are very lively, and move with a wriggling motion. They rest with the head downward, and with the hind extremity of the body — through which they breathe — at the surface of the water. At length they shed their skins and enter upon the pupa state, in which they live at the surface of the water, and breathe through two tubes on the thorax. In a few days the skin splits on the back, the winged insect appears, and, after resting awhile on its empty skin as it floats upon the water, spreads its wings, and, humming its war-note, flies away in search of a victim whom it may pierce for blood.

HESSIAN FLY AND WHEAT FLY.

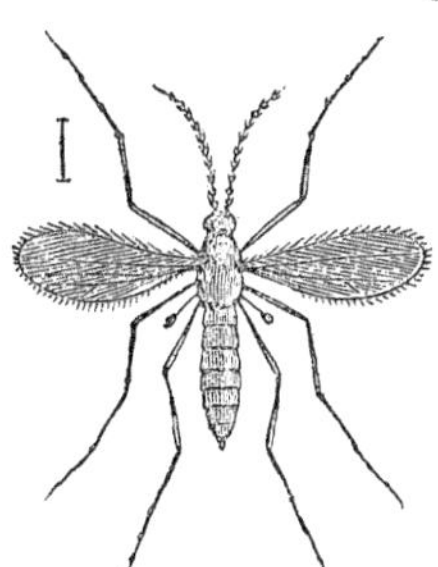

Fig. 291.—Hessian Fly.

The Hessian Fly expands only about one fourth of an inch, and has the head, antennæ, and thorax black, the wings blackish and fringed with short hairs. The hind body is tawny, with black on each ring, the legs brownish, and feet black. Two broods appear in a year, — one in spring and one in autumn. The females lay their eggs on the young blades of wheat, both in spring and fall. The eggs are only about one fiftieth of an inch in length, pale red, and they hatch in about four days, producing pale red maggots. The larvæ immediately crawl down the leaf till they come to a joint. Here they rest a little below the surface of the ground till they have undergone their transformations. They injure the plant by sucking its sap. The larvæ reach their growth in five or six weeks, and are then covered with a hardening brown or chestnut-colored skin, and the insect is then said to be in the flax-seed state, from its resemblance to a flax-seed. In April and May they complete their transformations, and come forth in the winged state, and soon begin to lay their eggs upon the spring wheat, and upon that sown the autumn before. The maggots hatched from these eggs pass down the stem as before stated, take the flax-seed form in June or July, and in autumn most of them are transformed into winged insects; others remain through the winter, and are transformed in the spring, as before stated. These flies sometimes move in immense swarms in search of fields of their favorite grain where they may lay their eggs. The Hessian Fly received its name from the belief that it was brought to this country in straw by the Hessian troops under the command of Sir William Howe.

The American Wheat Fly is about one tenth of an inch long, orange-colored, wings transparent, eyes black and prominent; antennæ long and blackish, those of the male being twice as long as the body, and consisting of twenty-four joints, and those of the female about as long as the body, and consisting of twelve joints. The wheat insects, in their perfect form, appear between the first of June and the last of August. They often move in immense swarms, taking wing in the morning and evening, and in cloudy weather, at which times they lay their eggs in the opening flowers of the grain, of barley, rye, and oats, as well as wheat. The eggs hatch in about eight days, producing little yellow maggots, which are found within the chaffy scales of the grain. The eggs are laid at different times, so that all do not come to maturity together; but they appear to come to their growth in about fourteen days. They prey upon wheat in blossom and in the milk, and do not touch the kernel after it has become hard. At length they cease eating, and soon after shed their skins, after which they become active again, and in a few days descend to the ground. Here they burrow, remain through the winter in the larva form, pass into the pupa state in early summer, and in a few days afterwards come forth in the winged state.

HORSE-FLIES.

These are generally large flies, having a proboscis enclosing very sharp lancets, with which they readily pierce the flesh of horses and cattle, in order to suck their blood. They have very large eyes, occupying nearly the whole head. There are several species, and some of the largest are nearly an inch long. The larvæ live in the ground. Figure 287.

ASILUS-FLIES.

These are very long-bodied flies, and are covered with stiff hairs. They are very rapacious, seizing and bearing away other insects. In the larva state they live in the roots of plants. One kind feeds upon the roots of the Pie Plant, or Rhubarb, of the gardens. Figure 289 shows a common kind of Asilus.

BEE-FLIES.

These flies are so named from their general resemblance to bees. They have a very long proboscis. They frequent sunny places in the woods, in the spring, and fly swiftly, but stop every little while and balance themselves in one place in the air.

BOT-FLIES.

These flies, in the larva state, live in various parts of the body of the ox, horse, and sheep, and occasion great suffering, and sometimes death, to these useful animals. One kind of Bot-Fly lays her eggs on the fore legs of the horse, another upon the lips, another upon the neck; by biting the parts, the horse gets the eggs into his mouth, swallows them, and the young hatch and cling to the walls of the stomach. The Ox Bot-Fly lays its eggs on the backs of cattle, and the larvæ live in burrows in the skin. The Sheep Bot-Fly lays its eggs in the nostrils of the sheep, and the larvæ crawl into the head, and often cause the death of the animal. Figure 290 is the Horse Bot-Fly.

BEETLES, OR COLEOPTERS.

Beetles are insects whose forward or upper wings are hard and horn-like, and meet in a straight line along the top of the back; and there is generally a little triangular piece between the bases of the wings, called the scutellum. The hind, or under wings, are thin, and when the insect is not flying are folded and concealed by the horn-like upper wings. The colors of beetles are often exceedingly beautiful and brilliant, rivalling even those of precious stones and the plumage of birds.

Beetles have two pairs of jaws, which move sidewise, by means of which they bite their food, which in some cases consists of other insects, in others of leaves or other parts of plants. In the larva state beetles are called grubs. The kinds are very numerous, probably not less than a hundred thousand in all.

TIGER BEETLES.

These are very common in warm sandy places, and may be seen in the roads in the country every pleas-

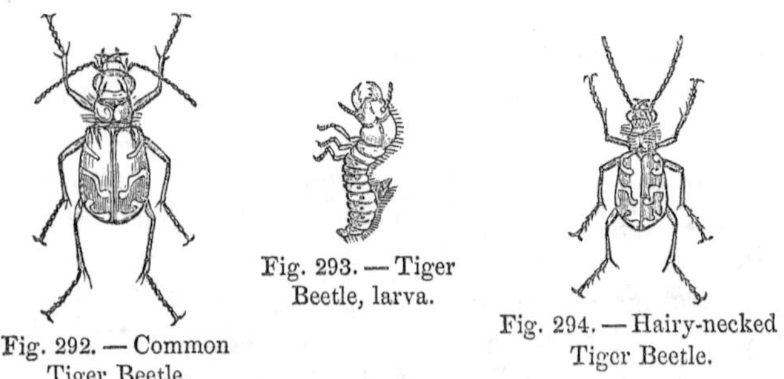

Fig. 292. — Common Tiger Beetle.

Fig. 293. — Tiger Beetle, larva.

Fig. 294. — Hairy-necked Tiger Beetle.

ant day. They are very beautifully and often splendidly colored, and have a large head and large eyes, and toothed jaws. They run rapidly, and fly when approached, but soon alight again. They devour great numbers of other insects for food, thus benefiting the farmer and gardener. The larvæ, or grubs, are soft, white, and are furnished with jaws like the adults; and, like the latter, feed on other insects, which they secure by digging holes in the ground, in which they remain, the head just closing the opening of the hole; and when some insect comes near enough, they seize it, draw it into the hole, and devour it.

GROUND BEETLES, OR CARABIDS.

These also prey upon other insects, and the kinds are very numerous. They have the jaws very long and hooked, and very long legs. Some of them have no under wings. One kind is called the Caterpillar

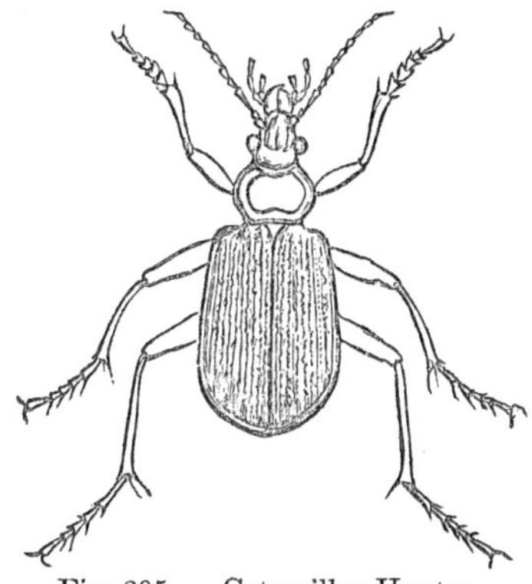

Fig. 295. — Caterpillar Hunter.

Hunter, because it destroys so many larvæ of other insects. It eats great numbers of the canker-worm, the most destructive insect which has appeared upon our beautiful fruit and shade trees, and which is described on page 153. It appears about the time the canker-worms leave the trees and come to the ground. The Glowing Caterpillar Hunter is a smaller kind, and is black, with six rows of sunken brilliant red spots.

WATER BEETLES.

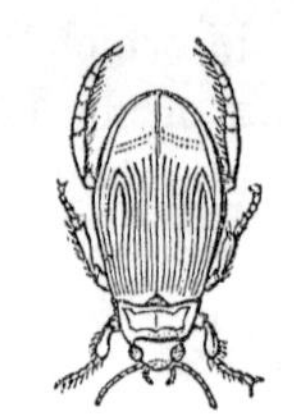
Fig. 296. — Water Beetle.

These beetles live in the water, and their long hind legs are well fitted for swimming, being fringed on their inner side. They are very voracious, and devour other insects, and, in some cases, young fishes. Some of the species are more than an inch long. The Whirligig Beetles which are found on the surface of still waters, where they look like brilliant spots gliding in all sorts of curves, are much smaller, and belong to another family.

CARRION BEETLES.

Carrion Beetles live together in great numbers in the bodies of decaying animals. Some kinds have the habit of burying the small animals which they find dead, and it is remarkable how quickly they find out where such animals are. If a dead frog, or mouse, or bird be placed upon the ground, these beetles will be seen about it in a few hours; and beginning to dig beneath it, they soon sink it out of sight. The females then lay their eggs in it, so that when the young hatch they find themselves amidst a supply of suitable food.

Fig. 297. — Carrion Beetle.

ROVE BEETLES.

These are long and narrow, with stout jaws, and the hind body much longer than the wing-covers. When they run they raise the hind body and move it in different directions. They are found about decaying substances. The larvæ closely resemble the perfect insect.

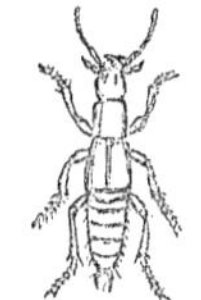
Fig. 298.—Rove Beetle.

HORN-BUGS.

Horn-Bugs are beetles which have the body very hard and oblong, the thorax and head very large, and the upper jaws large and often curved and branched.

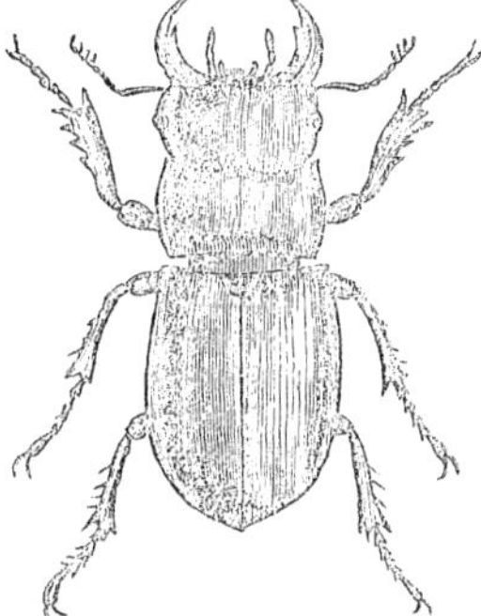
Fig. 299.—Horn-Bug.

They keep in their hiding-places in the daytime, and fly about at night. In the adult state they eat the leaves of trees; but the grubs live in the trunks and roots of trees, and some kinds thus live in the larva state for six years before they become perfect beetles.

SCARABÆIANS.

The beetles known as Scarabæians have the antennæ ending in a knob, which is made up of three or more leaf-shaped pieces, and they have a sort of plate which extends forward over the face like the visor of a boy's cap, and their legs are toothed on the outer sides, and thus fitted for digging. Some live on the ground and are called Ground-Beetles; others live upon trees, whose leaves they eat, and are called Tree-Beetles; others feed upon the sweets of flowers, and are called Flower-Beetles. Some kinds are very large, as the

Fig. 300.—Goldsmith Beetle.

Fig. 301.—Phanæus.

Hercules Beetles of South America, which are five inches long. Many are brilliantly colored, and the Phanæus has a horn-like projection on the head. The May-Beetles are brown-colored Scarabæians, which, attracted by the light, fly into our rooms in the early part of summer; in the grub state they live in the ground, and are white, with a brownish head. The Goldsmith Beetle is of a beautiful golden color above, and copper color, with whitish wool, below. It feeds upon leaves, among which it hides by day, flying in the morning and evening twilight. The Spotted Pelidnota is found on the grape-vine in July and August. It is about an inch long, brownish yellow above, with three black dots on each wing-cover, and one on each side of the thorax.

Many of these beetles not only injure the foliage of

shrubs and trees, but in their grub or larva state they devour the roots of grasses and other plants, and thus do immense injury to the crops. Fortunately, however, they are kept in check by the crow and many other animals, which eagerly devour them.

BUPRESTIANS.

These beetles, in the larva state, live in the trunks of trees, eating holes in all directions, much to the injury of the tree. Some kinds bore the peach, others the plum, others the oak, and others the pine. The perfect beetles are long and very solid, with a sunken head, and often with metallic colors.

Fig. 302. — Buprestis.

SPRING OR SNAP BEETLES, OR ELATERS.

When placed upon the back, these beetles at once, with a snap and a jerk, throw themselves upwards; and they repeat the operation till they come down right side up. They perform this feat by means of a spine-like organ situated on the under side of the breast. Snap-Beetles vary from half an inch to two inches in length, and the head is almost concealed in the thorax. One of the most curious kinds has two eye-like spots on the thorax, as seen in Figure 303.

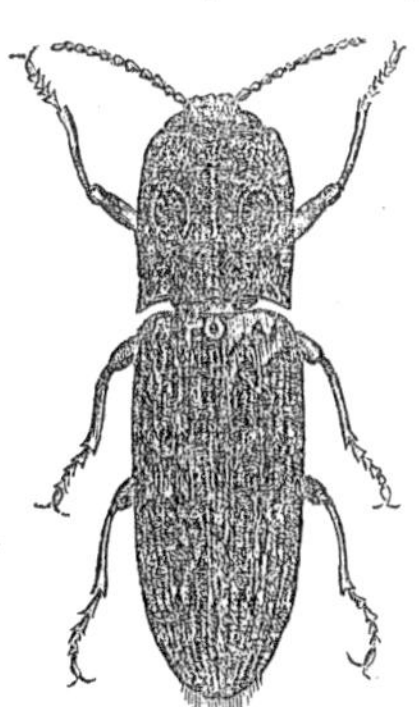

Fig. 303. — Eyed Spring Beetle.

Some of the Elaters, and others closely related to them, give out a brilliant light at night, and are known as Fire-Flies. They are common in meadows in summer. Some of the tropical kinds emit such a brilliant phosphorescence — as their light is called — that a few of them placed in a glass vessel give light enough for a person to read by.

CURCULIOS, OR WEEVILS.

These beetles are hard, generally rather small, some being minute, and in most cases they have a long, slender snout. In some, however, the fore part of the head is broad. They feign death when disturbed, and,

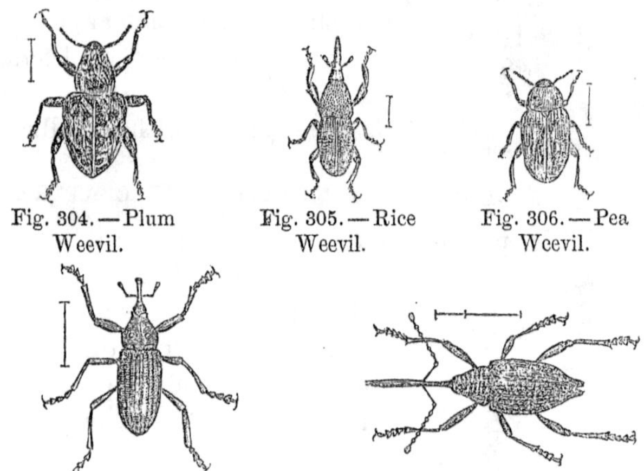

Fig. 304. — Plum Weevil. Fig. 305. — Rice Weevil. Fig. 306. — Pea Weevil.

Fig. 307. — White-Pine Weevil. Fig. 308. — Long-snouted Nut Weevil.

if upon a tree, fall to the ground and remain motionless till all is quiet. The Pea Weevil lays its eggs on the pea blossoms, and the grub enters the pea through the green pod, and remains there till the next spring, when it comes out as a perfect beetle or weevil. The Baltimore Oriole splits open the pods for the sake of obtaining the grubs contained in the peas. The White-

Pine Weevil, in the larva state, lives in the trunk of the pine, in which it cuts passages in various directions. The Long-snouted Nut Weevil, in the larva state, lives in nuts. The Plum Weevil, when shaken from the tree, looks like a dried bud. This weevil makes a crescent-shaped wound on the surface of the plum, in which it lays an egg; from the egg there hatches a whitish grub, which burrows into the plum, even to the stone. The Rice Weevil feeds upon rice, wheat, and Indian corn. It is about one tenth of an inch long, with two red spots on each wing-cover.

LONG-HORN, OR CAPRICORN BEETLES.

These beetles have very long and generally curved antennæ. When we catch them they make a squeaking

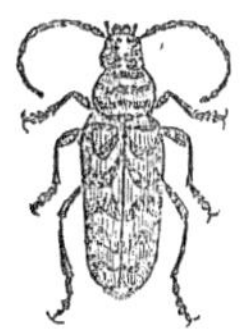

Fig. 309. — Painted Clytus.

Fig. 310. — Apple-tree Borer, larva.

Fig. 311. — Apple-tree Borer, adult.

sound, by rubbing together the joints of the thorax and hind body. In the larva state they live in the trunks of trees and in timber, and are called *borers*. As they eat their way in the timber they fill the passages behind them with their cuttings, which the carpenters call *powder-post*. Some, however, as the Apple-tree Borer, keep the end of their burrow open, out of which they cast their chips. They remain in the larva state from one to three years.

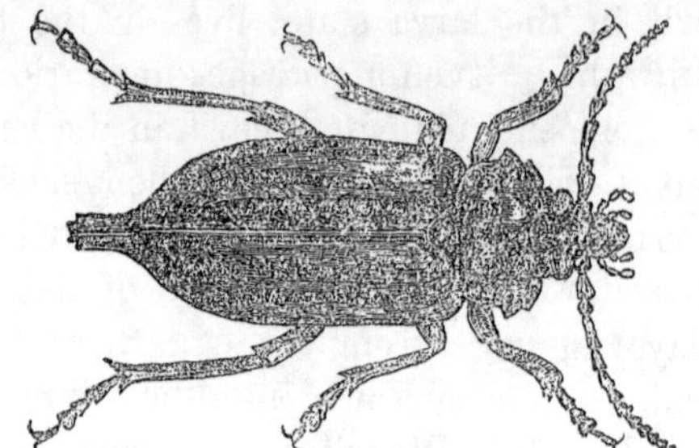

Fig. 312. — Broad-necked Prionus, a Capricorn Beetle.

CHRYSOMELANS AND LADY-BIRDS.

These are either egg-shaped or hemispherical, and are remarkable for their beautiful colors. The first are

Fig. 313. — Ladder Chrysomelan.

Fig. 314. — Cucumber Beetle, a Chrysomelan.

Fig. 315. — Lady-Bird.

blue, green, or golden; the latter are black, red, or yellow, with dark spots. The Lady-Birds devour plant-lice, and are thus of great benefit to the gardener.

BUGS, CICADAS, AND TREE-HOPPERS, OR HEMIPTERS.

These insects have a slender, horny beak, which, when not in use, is bent under the body and lies upon the breast.

CICADAS, OR HARVEST FLIES.

The Cicadas, or Harvest Flies, have a very large head, large eyes, and three minute eyes on the top of the head, and their wings are large, thin, and very distinctly veined. The males make a very loud buzzing sound by means of curious organs resembling kettle-drums, one being placed on each side of the hind body near the thorax. The ancient Greeks loved to hear the buzzing of the Cicadas, and kept them in cages that they might enjoy their rude music. These people also ate Cicadas. The females have a very curious piercer for making holes in trees, in which to lay their eggs. This piercer consists of three pieces, the two outer ones grooved on the inside and toothed on the outside like a saw, and a central borer which plays in the groove formed by the other two.

Fig. 316. — Seventeen-year Cicada.

The Seventeen-year Cicada is about an inch long, the general color black, with the eyes, larger veins, and

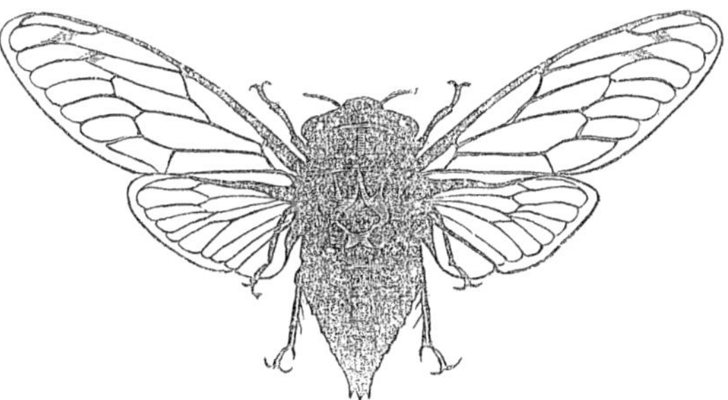

Fig. 317. — Dog-day Cicada, or Harvest Fly.

forward edges of the wings red. This is generally

called the Seventeen-year *Locust*, but it is in no sense a Locust, and should not be called by this name. The name "Seventeen-year" is given to it from the belief that it appears in the same place only once in seventeen years.

The Dog-day Harvest Fly is over an inch long, the body black above, marked with green, and the under side covered with a white substance resembling flour. It appears at the beginning of dog-days, and its singing may be heard among the trees through the middle of the day. The pupæ of this species and of the Seventeen-year Cicada, as they come out of the ground and crawl up the trees, look like beetles. Soon the pupa-skin splits on the top of the back, and from the opening thus made the perfect Cicada comes forth, leaving the brown pupa-skin attached firmly to the tree, and at a little distance looking as when alive.

TREE-HOPPERS.

These insects are remarkable for their curious and

Fig. 318. — Tree-Hopper.

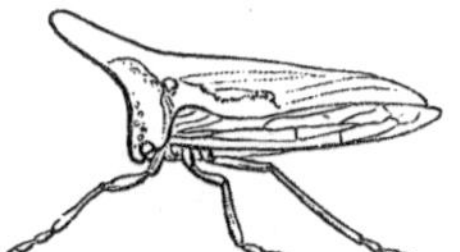

Fig. 319. — Same enlarged.

often grotesque shapes. They live on the sap of trees and herbs, and imbibe it in such quantities that it oozes out of the body, often concealing the insect in a mass of frothy matter or foam. Figure 318 shows one of the common kinds, as seen when looking upon its back, and Figure 319 is the same in profile, and considerably enlarged.

8

APHIDES, OR PLANT-LICE.

These insects have the body short, and at the hind extremity there are two little tubes, from which come minute drops of a very sweet fluid. Aphides inhabit all kinds of plants, the leaves and softer portions being often completely covered with them. The young are hatched in the spring, and soon come to maturity, and, what is remarkable, the whole brood consists of wingless females; and what is still more remarkable, these females bring forth living young, each female producing fifteen or twenty in a day. These young are also wingless females, and at maturity bring forth living young, which are also all wingless females, and in their turn bring forth living young; and in this way brood after brood is produced, even to the fourteenth generation, in a single season. But the last brood in autumn contains both males and females, which stock the plants with eggs, and then perish. Réaumur, a celebrated naturalist, has proved that a single aphis, in five generations, may have about six thousand millions of descendants! Wherever plant-lice abound, ants collect to feed upon the honey-like fluid produced by them; and the most friendly relations exist between these two kinds of insects. An aphis has been known to give in succession a drop of the fluid to each of a number of ants waiting to receive it!

Fig. 320. — Aphis.

SCORPION BUGS.

These bugs live in the water, and can sting severely. They devour other insects, which they seize with their fore legs, which act as pincers.

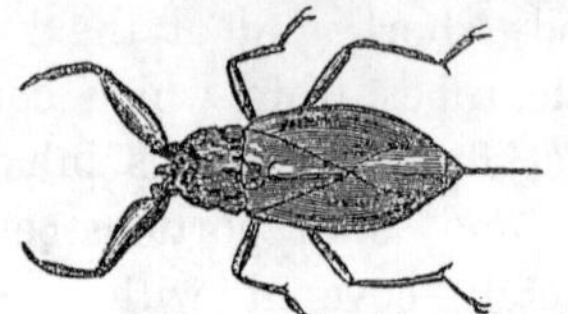
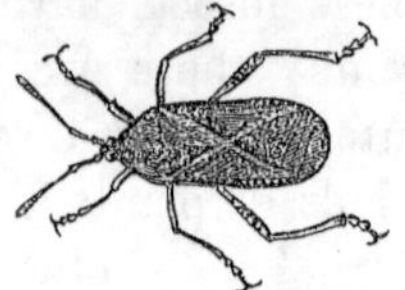
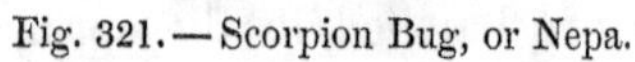

Fig. 321.—Scorpion Bug, or Nepa. Fig. 322.—Squash Bug.

SQUASH BUGS.

The Squash Bug passes the winter in a torpid state, and when the leaves of the squash appear it lays its eggs in clusters on the under side of them.

STRAIGHT-WINGED INSECTS, OR ORTHOPTERS.

These insects have wings which lie straight along the top or sides of the back. They do not pass through the marked stages of larva and pupa in coming to the adult state; but the young are constantly active, feeding and growing, and differ from the adults only in size, and in having only rudiments of wings, and in frequently changing their skins. After having shed their skins six times, they come forth perfect insects.

EARWIGS.

These insects have a pair of sharp-pointed nippers

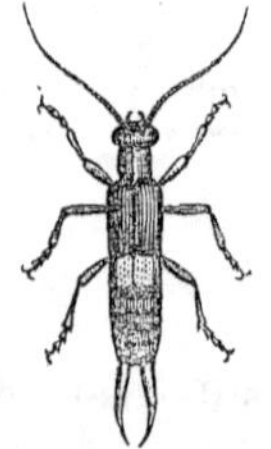

Fig. 323.—Earwig.

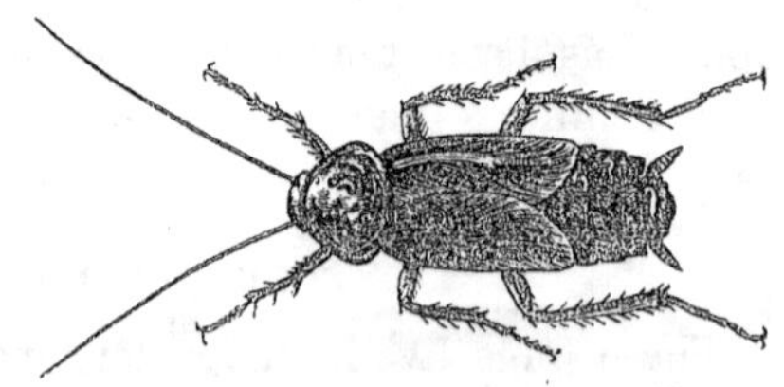

Fig. 324.—Cockroach.

at the hind part of the body, which they can open and

shut like a pair of scissors. They are found under stones, and under the bark of old trees, and fly only at night. They are believed by some to crawl into people's ears, but there is probably no good reason for this belief.

COCKROACHES.

Cockroaches are found in forests, and some species infest kitchens, store-rooms, and closets, devouring all kinds of food, and even clothes. Figure 324 shows one of the kinds common in this country, although it originated in Asia.

WALKING-STICKS AND WALKING-LEAVES.

The Walking-Sticks are insects which look like dry twigs; and the Walking-Leaves have wings that look almost precisely like leaves. They belong mostly to

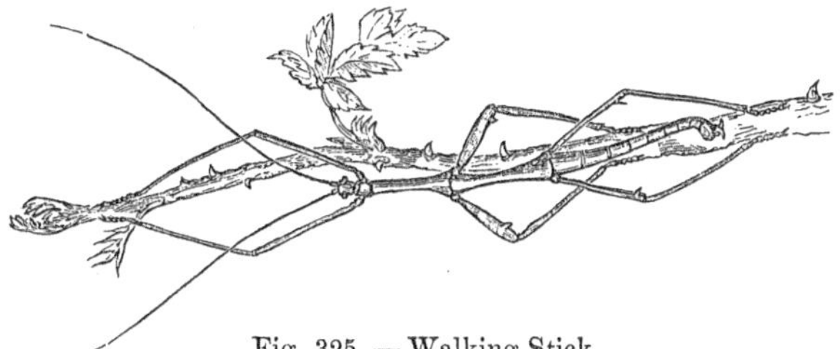

Fig. 325. — Walking-Stick.

the warm parts of the globe, but several kinds of Walking-Sticks are found in the United States. They are from three or four inches to a foot long. Figure 325 shows one of our most common species, about one half its natural size.

MANTES.

The Mantes are grasshopper-like insects which have the fore legs suited for seizing and holding prey. They are found upon plants and trees, where they sit for hours, holding up their fore legs, ready to seize any

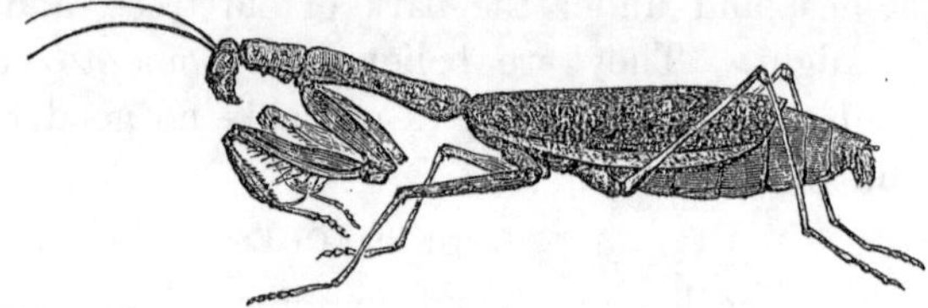

Fig. 326.—American Mantis.

insect which comes within reach. Some of the superstitious inhabitants of the East believe that at such times the Mantis is engaged in religious devotions. Figure 326 represents the only kind found in the United States.

CRICKETS.

Crickets have a flattened body, long antennæ, and long appendages behind. The males chirrup to attract their mates, and this familiar sound is often heard throughout the night. It is produced by rubbing the wings against one another. The most common crickets of the fields are dark-colored, but some, like the Climbing Crickets, are white. The Mole Crickets have fore feet resembling those of the Mole, and well adapted for digging. They burrow in the ground, and prey upon other insects. Some kinds of crickets take up their abode in houses, and the sound of "the cricket on the hearth" is a familiar one to people who live in the country.

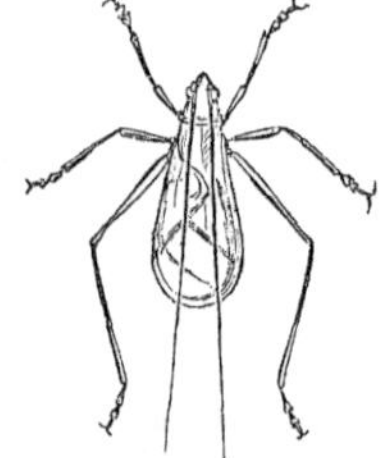

Fig. 327.—White Climbing Cricket.

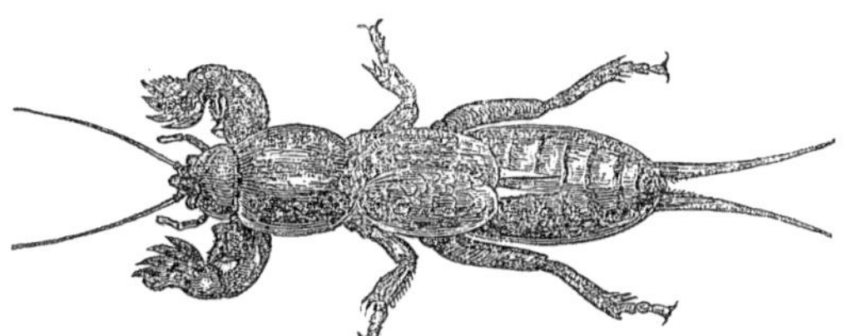

Fig. 328. — Mole Cricket.

LOCUSTS.

These are grasshopper-like insects which have very long antennæ, a long ovipositor, and many of them

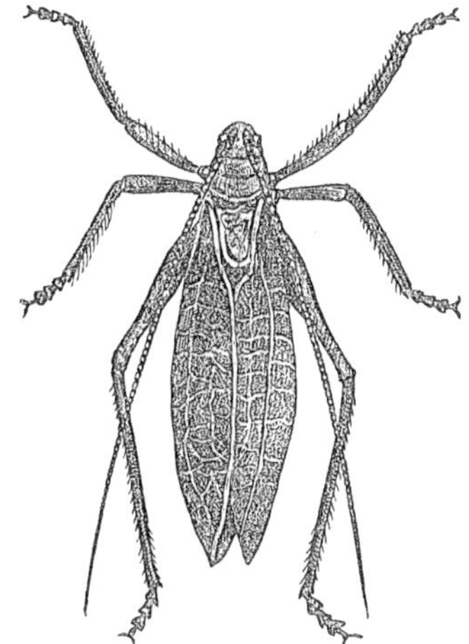

Fig. 329. — Katydid.

produce a grating noise by rubbing their wing-covers together. They are known as Katydids, Leaf-winged Grasshoppers, Sword-bearers, and Meadow Grasshoppers. None are more interesting than the Katydids, whose curious notes are heard at early twilight or on moonlight evenings, and in cloudy days, throughout the autumn. These insects are about an inch and a half long, and the wings shut around the body like the two valves of a pea-pod. They produce sounds resembling the words "Katy did." These are made by means of a thin membrane stretched in a strong frame which is situated in the overlapping portion of each wing-cover. The rubbing of the frames against each other, as the insect opens and shuts its wings, makes the sounds.

MIGRATORY LOCUSTS.

These are grasshopper-like insects which have the antennæ short, and no long organ for laying eggs. The kinds are many, and some of the tropical ones are

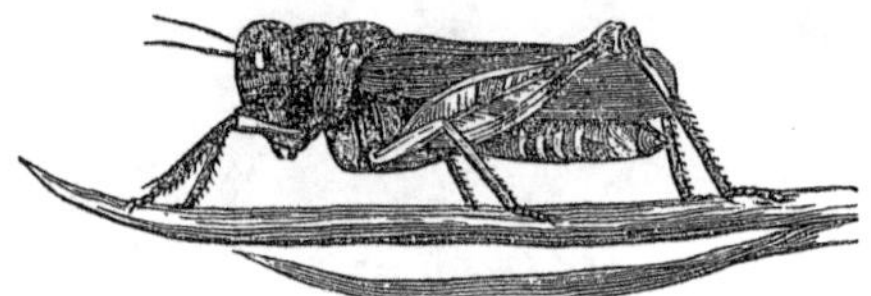

Fig. 330. — Clouded Locust.

three or four inches long. The most common grasshoppers of the United States belong in this group, and are familiarly known as the Red-legged Locust, Carolina Locust, Coral-winged Locust, Yellow-winged Locust, and Clouded Locust.

NET-WINGED INSECTS, OR NEUROPTERS.

These insects have four thin, finely net-veined wings, stout jaws, and no sting. In the larva state they live in the water. The Stone-Flies, Figure 331, and Ephemeras, Figure 332, have long appendages at the hind part of the body.

MAY-FLIES, OR EPHEMERAS.

Though these insects live only for a few hours or a day in the perfect state, their existence in the larva and semi-pupa state extends through two or three years,

Fig. 331. — Stone-Fly, half natural size.

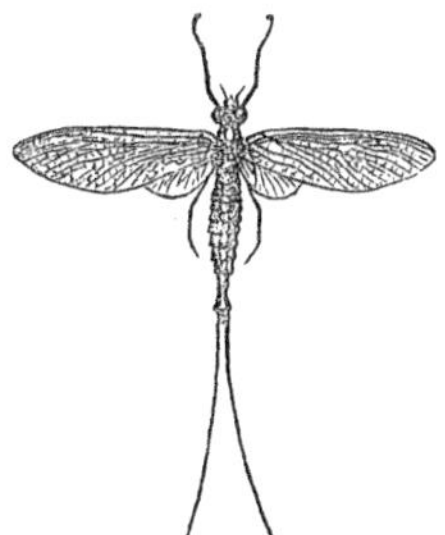

Fig. 332. — May-Fly.

and all this time they live in the water. When about to go through their final changes, the pupæ crawl to the surface, cast off the pupa-skin, and appear at first to be fully developed; this is the sub-imago state; they then fly with difficulty to the shore, affix themselves to plants and trees, and cast off a very delicate covering. After this the wings are brighter, and the tails greatly increase in length. May-Flies appear in such immense swarms in some parts of Europe, that the people collect their dead bodies into heaps to enrich the land. They are common in this country. One of our species is shown in Figure 332.

DRAGON-FLIES, OR DARNING-NEEDLES.

These insects have a long body, large, lustrous, gauze-like wings, large head, and very large eyes. They at once arrest our attention by their large size, light and graceful form, variegated colors, and the great velocity with which they speed their way over fields and meadows, or skim the surfaces of the pools or ponds in

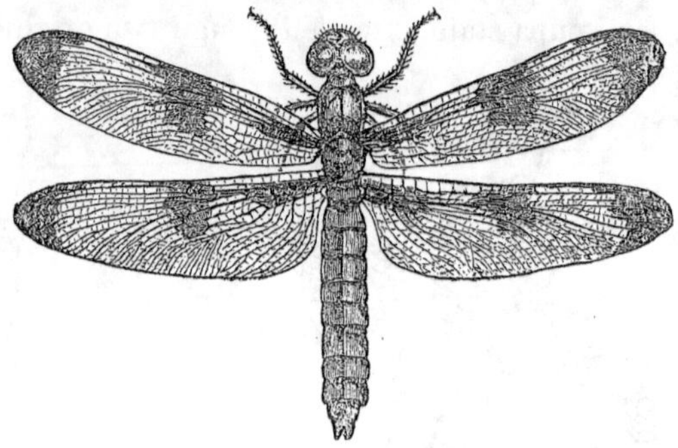

Fig. 333.—Dragon-Fly.

search of flies, mosquitoes, and other insects, upon which they feed. In the larva and pupa states they live in the water, and are rather long, broad, and flat, with long sprawling legs, and they crawl about, or propel themselves by ejecting water from a cavity situated at the hind part of their body. They are very voracious, devouring other insects and even one another. When the time comes for the last change, they crawl up the stems of plants, and, having withdrawn from the pupa-skin, which remains clinging to the plant, and dried themselves a little, they spread their wings and dart swiftly away. Though they bite quite fiercely with their jaws, they are without any sort of sting, and are perfectly harmless to man.

CORYDALIS.

The Horned Corydalis expands five or six inches, and the male has two long, horn-like pincers.

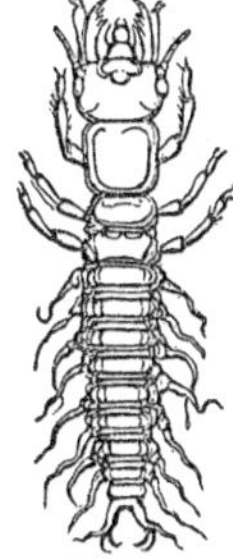

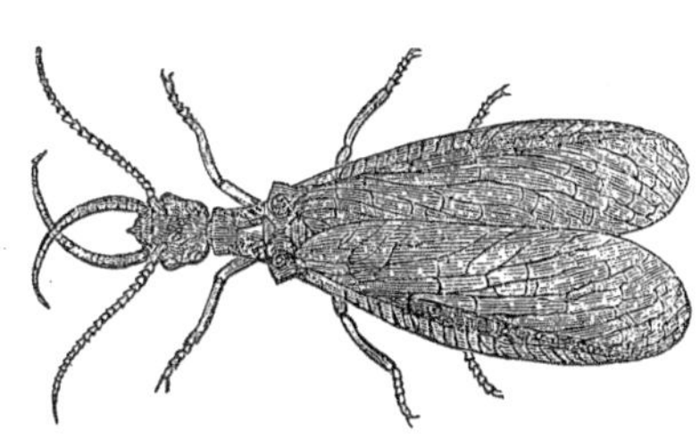

Fig. 334. Fig. 335.
Larva and Perfect Insect of Horned Corydalis, one half the natural size.

LACE-WINGS, OR ANT-LIONS.

The Ant-Lion is so called because, in the larva state, it preys upon ants and other insects, which it secures in the following manner: it makes a pitfall, or cavity,

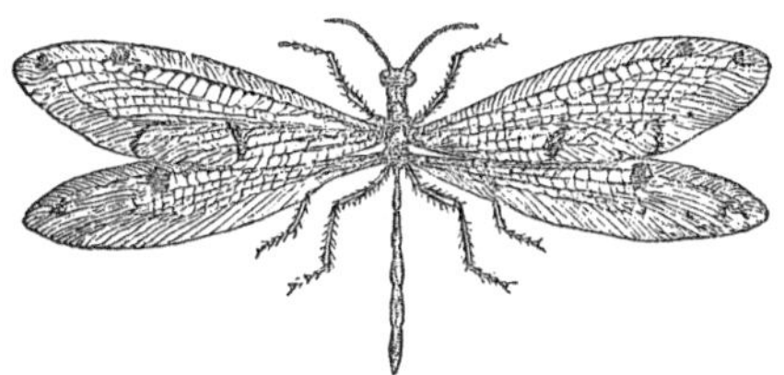

Fig. 336.—Ant-Lion.

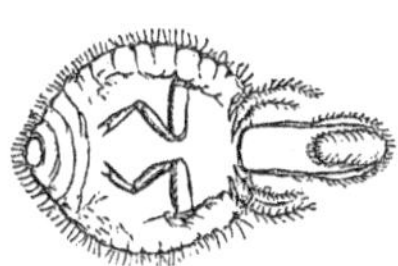

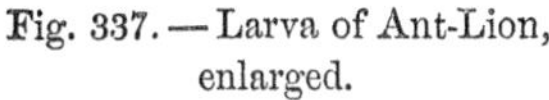

Fig. 337.—Larva of Ant-Lion, enlarged.

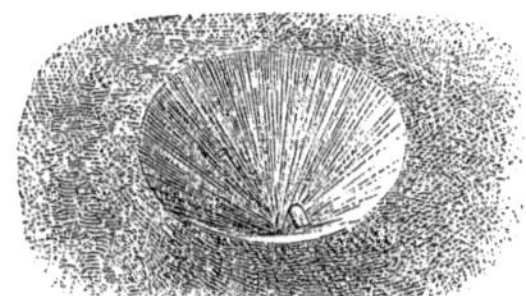

Fig. 338.—Pitfall of Ant-Lion.

Figure 338, at the bottom of which it conceals itself, excepting its jaws, and there awaits its prey. Whenever an insect falls into the pit, the ant-lion rushes upon it and devours it.

CADDICE-FLIES.

On account of their curious appearance and habits, these insects are the most interesting while in the larva state. They live at the bottom of ponds and streams,

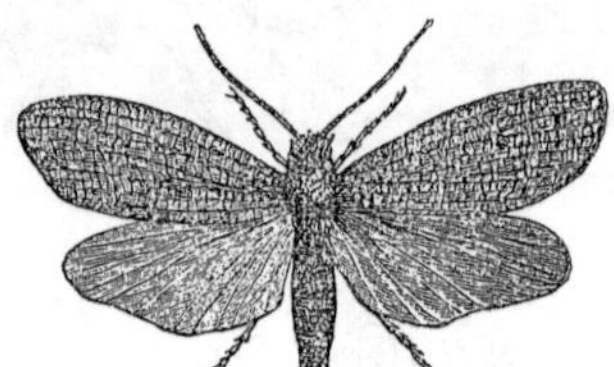

Fig. 339. — Caddice-Fly.

in cases which they construct of bits of wood, or grasses, or of grains of sand, or of fragments of broken shells, and which are lined with silk, which they spin from their mouths. They sometimes load one side of the case with heavier pieces, in order to keep that side downward.

SPIDERS, OR ARACHNIDS.

Spiders have the body divided into only two well-marked portions, — the head and the hind body. They have eight legs, and two palpi or feelers resembling legs, but no wings, and they do not change their form in passing from the young to the adult state. Most kinds feed upon insects.

True Spiders have, at the hind part of their body, a most wonderful organ, called the spinneret, by which the delicate threads of the spider-web are spun. It consists of four to six knobs, with a thousand or more holes in each knob. Through these the invisible silken threads pass out, — more than four thousand at a time, — and at a little distance from the knobs all these unite into one, forming the single line of spider-web which all are so familiar with. As the threads issue from the knobs they are a sticky fluid,—which has

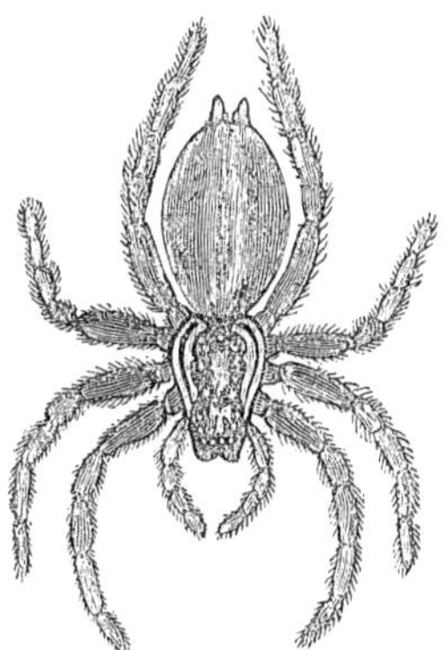

Fig. 340.—Spider—Lycosa.

been secreted in little bags in the abdomen: but this hardens into silk as soon as it comes to the air. The length of the line which a spider is able to produce is truly wonderful. Dr. Wilder wound nearly two miles of silk, in less than a day, from his celebrated *Nephila plumipes*,—a spider which he discovered in South Carolina. The kinds of Spider are very numerous, and most of them spin some sort of a net-like web, in or near which they live, and by means of which they capture insects for food. The House Spider spreads a flat net in the corners of rooms. The Geometric Spider spreads a vertical net, which is made in the most beautiful manner, radiating lines running from the centre, like the spokes of a wheel, and these connected by a spiral line, which at a little distance gives the appearance of lines arranged in circles from the centre outwards. Some kinds of Spider have, near the principal web, a silken retreat, or den, where the owner hides till the quivering spider-lines which run into its office telegraph the fact that a fly has become entangled; instantly the spider rushes out of its retreat, pounces upon the victim, and bites it, if possible, putting into the wound a fatal poison. If the insect be too powerful for the spider, the latter waits till the insect gets more entangled, and finally exhausted, by its efforts to escape, then binds it with silken bands, and begins to devour it. The bite of an ordinary spider will kill a fly; the bite of some of the large kinds in South America kills the humming-bird; and sometimes men are killed by a spider's bite. The female spiders lay eggs and enclose them in silken sacs. Some kinds carry the egg-sac about with them; others spin it in a safe place, and, in some instances, stay near to guard it, and to tear open the egg-sac as soon as the young are hatched, that they may escape. One of the most curious of these egg-sacs is that shown in Figure 341, and which was made by some spider which we do

Fig. 341. — Egg-case of a Spider, — the Vase-Maker.

not yet know, but which may properly be called the *Vase-Maker*. Two "vases," like the one in the woodcut, were found standing about a foot apart on the stem of a grape-vine. The outside of the vase looks like brown paper, or it is in appearance and in toughness like the outside of the cocoon of the Cecropia Moth, and the vase is fastened on to the vine by a vast number of threads of silk passing from one side of the vase to the other around the vine; and the threads are so nicely arranged that the vase cannot turn nor slip from its place. On opening this curious structure, it was found to be filled with the finest silk and a great number of newly-hatched spiders. This rare and wonderful specimen of spider-building and of spider-case for its young was presented to the Zoölogical Cabinet of Vassar College by S. M. Buckingham, Esq., of Poughkeepsie, New York.

SCORPIONS.

The Scorpions are confined to warm regions, and live among ruins of buildings, under rubbish, and sometimes in houses. They have a long body ending in a

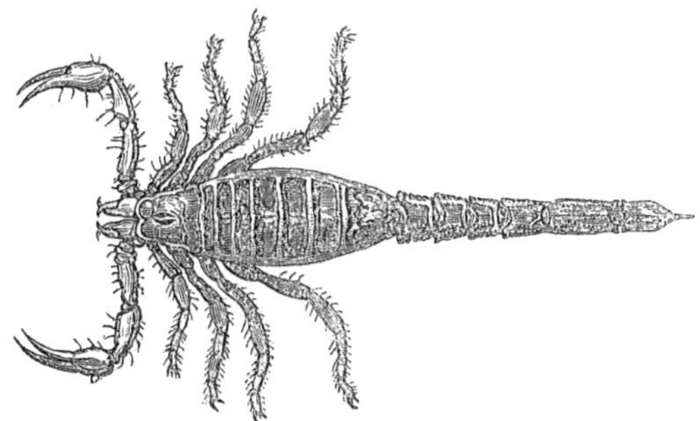

Fig. 342.—Scorpion.

curved, sharp sting, with which they inflict dangerous and sometimes fatal wounds. They can run quite rapidly, and can bend the hind body or tail in any direction, and use it both for attack and defence. The one here figured is found in Texas.

CENTIPEDES, OR MYRIAPODS.

These are very long and worm-like, and divided into very numerous rings or joints, each one of which generally bears two pairs of feet. In the temperate parts of the globe the kinds are not more than two or three

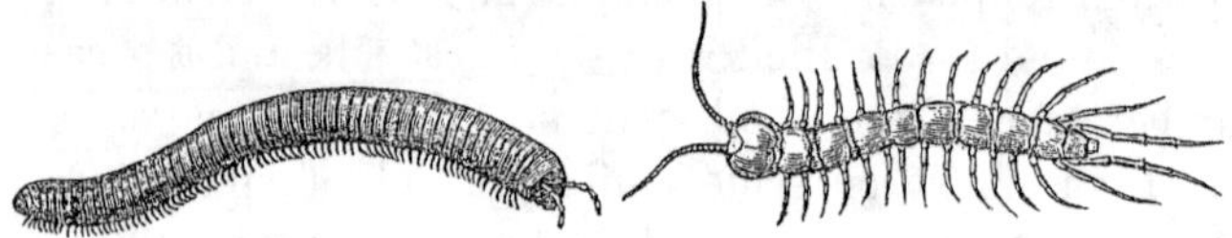

Fig. 343.—American Myriapod, or Galley-Worm.

Fig. 344.—American Earwig, or Lithobius.

inches long; but tropical species are a foot long in some instances, and the bite of these is often very poisonous. All prey upon insects.

CRUSTACEANS.

These articulates have a crust or shell, the head and thorax united into one piece, and they live in the water and breathe by means of gills. Some kinds, however, live upon the land. They feed upon all sorts of animal food, and shed and renew their shell many times.

TEN-FOOTED CRUSTACEANS, OR CRABS, LOBSTERS, AND SHRIMPS.

Crabs can walk forward, backward, and sidewise. The tail, or hind body, is small, and is doubled under the forward part of the body, where it fits into a groove. The kinds of Crab are very numerous, and some are found on every sea-coast. They vary in size from that of a penny to those which, with the legs outspread, cover a space a yard square. Some kinds are very much prized for food; the one shown in Figure 347 is sold in great numbers in the markets of New York and Philadelphia. Hermit Crabs have the hind

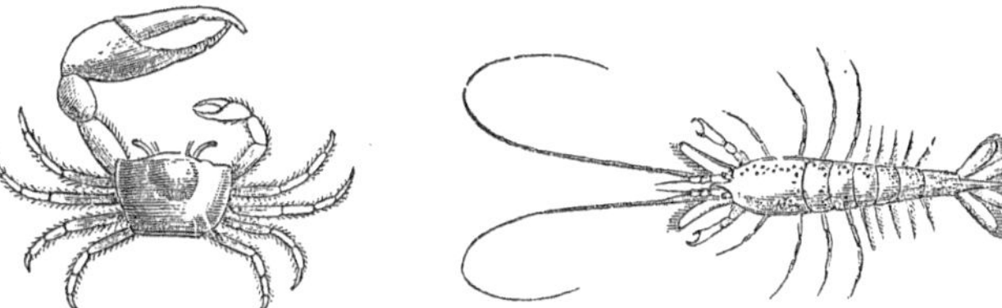

Fig. 345. — Fiddler Crab.

Fig. 346. — Bait Shrimp.

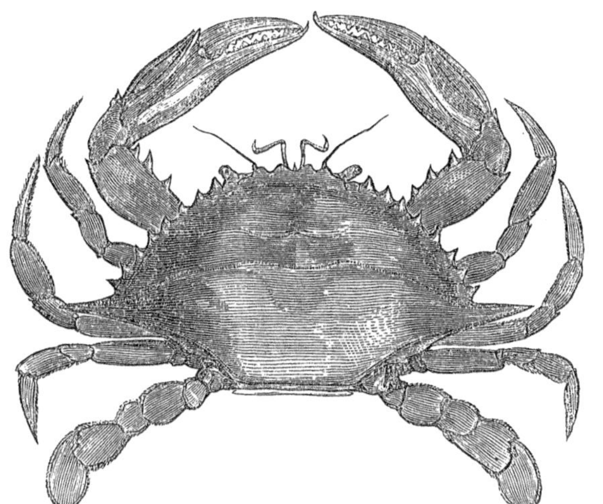

Fig. 347. — American Edible Crab.

part of the body long, soft, and tapering, and they take up their abode in empty univalve shells, which they drag about with them wherever they go, and they look as though they were the real and original owners of the houses which they live in. When a Hermit Crab becomes too large for the shell which it has chosen for its home, it abandons it, and begins its search for a new one, inserting itself backwards into one shell after another till one is found which suits it. When not moving about, or when alarmed, it retreats as far as possible into the shell, and closes the opening with its larger claw.

Lobsters and Shrimps have the hind body, often called the tail, large and long, and generally turned forward, as seen in Figure 348. The American Lobster is from one to two feet long, and weighs from three to ten pounds or more. It is very abundant on the coast of New England, and great numbers are caught in lobster-pots baited with fish, and are sold in the markets of Boston, New York, and other cities.

Two of the forward leg-like appendages of lobsters are greatly enlarged, and end in powerful claws or pincers. One of these is provided with blunt teeth, or tubercles, suited for crushing shells, and the other with exceedingly sharp teeth suited for biting. So powerful are these organs that with them a lobster can easily bite off a man's finger; and if one were to get hold of your hand, you could release it only by breaking off the lobster's claw. The fisherman, well knowing their biting powers and habits, puts a wooden plug into the joints of their pincers, so that they cannot open them; if this were not done, the lobsters, when confined in the lobster-car, — a large box in the water where lobsters are kept after they are caught, — would bite off the limbs of one another. In crawling the lobster moves rather slowly, but sometimes, by a single stroke of its powerful tail or hind body, it darts through the water, backwards, a distance of fifteen or twenty feet, with the swiftness of an arrow. When a lobster or other crustacean loses a leg or other organ, another like it grows to supply its place. But one of the most remarkable facts about lobsters and other crustaceans is, that from time to time they shed the shell in one piece, so that the cast-off shell looks exactly like the perfect animal, — antennæ, eyes, jaws, legs, paddles, and even every hair, are all just as they were when they covered the live lobster! The lobster

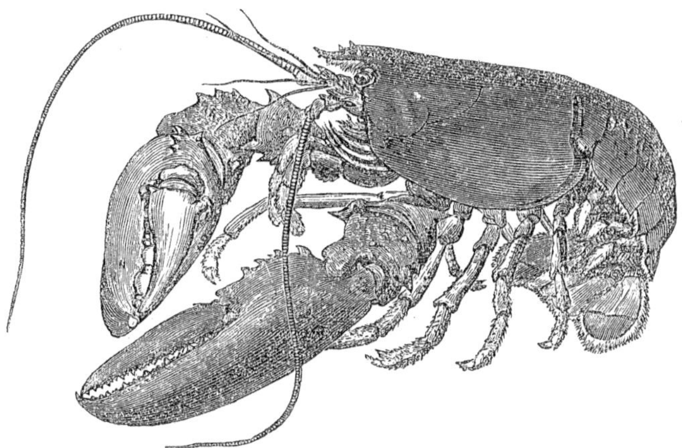

Fig. 348. — American Lobster.

comes out of its shell through a rent on the back, and is at first very soft; it at once increases in size, and in a few days its skin becomes as hard as the shell which it cast off. This shedding of the shell is necessary for the growth of these animals, for while the shell remains the lobster or other crustacean can grow only just large enough to completely fill it. When a lobster is ready to shed its shell, there are two hard, stone-like bodies at the sides of the stomach, and it is supposed that these furnish a part of the solid matter for the new shell; for they immediately begin to grow smaller after the moulting, and soon entirely disappear.

The Craw-Fish, or Fresh-water Lobster, much resembles the American Lobster, but is only three or four inches long, and lives in brooks. One kind is common on the Western prairies, where it lives in holes which it digs in the ground deep enough to find water.

SAND-FLEAS, &c., OR FOURTEEN-FOOTED CRUSTACEANS.

Beach- or Sand-Fleas are little shrimp-like crustaceans which are very common on the sea-beach. They

Fig. 349. — Sand Flea.

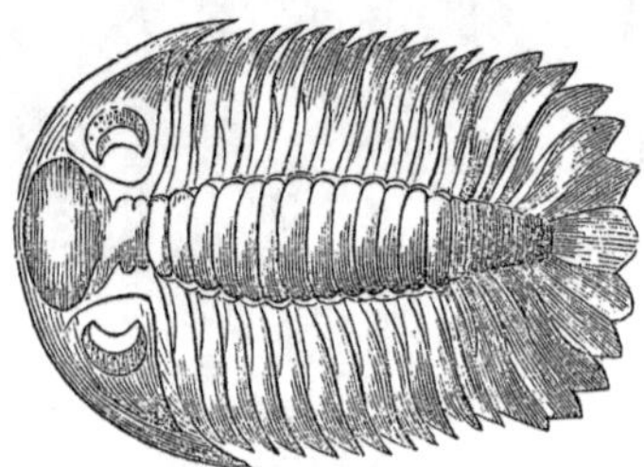

Fig. 350. — Trilobite.

have seven pairs of feet. Closely related to these are the curious Trilobites, found imbedded in the solid rock, and which lived and died ages ago.

BARNACLES AND HORSE-SHOE CRABS.

The Barnacles are of many kinds. Some resemble bivalve shells, and grow in clusters, attached by stems, as seen in Figure 351; others, as in Figure 352, are acorn-shaped, and are fixed directly upon the rocks, shells, lobsters, or ship-bottoms. They are all provided with feather-like arms or feet, which they regularly protrude and withdraw, — a sort of grasping motion as though they would secure any little animals or particles of food that might be within their reach. Some kinds of Acorn-Barnacle completely cover the rocks between high and low water mark; others delight in deep water. In long voyages barnacles sometimes become so numerous on the bottom of a vessel as to seriously hinder its progress. Although in the adult state Barnacles or Cirripeds are fixed and stationary, the young swim freely about.

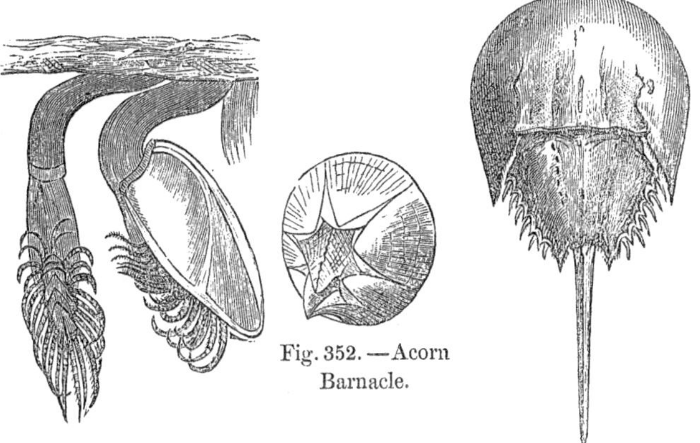

Fig. 352. — Acorn Barnacle.

Fig. 351. — Duck Barnacle.

Fig. 353. — Horse-shoe Crab.

The Horse-shoe Crab is found in all seas. Some are two feet in length, and in all cases the body ends in a sharp spine, which some of the savage tribes use for spear-points. This curious crab walks and eats with the same organs! — the lower part of the first six pairs of legs being used for walking, and the upper parts of the same legs being provided with teeth-like organs, and used for jaws.

WORMS.

Worms are long animals, which are made up of many similar rings. The nerves are distributed in knots or clusters throughout the whole length of the animal, and hence worms are not killed when cut in pieces; and in some cases the several pieces become distinct worms. The kinds of worm are very numerous, but they are most abundant in the sea and in fresh waters. Most of the animals which look like worms are butterflies, or moths, in the larva state, and are caterpillars instead of worms. One of the most common kinds of sea-worm is the Serpula, which lives in tubes that are found incrusting stones and other bodies. The breathing organs are in tufts near the head, and there is a little round body, shown in the cut, which serves to shut the animal in when it withdraws itself into the tube. The Angle- or Earth-Worm, common in rich soils, is well known to all boys, and is used as good bait for trout and other fishes. One of the most

Fig. 354. — Serpula.

Fig. 355. — Earth-Worm.

common of the fresh-water worms is the Gordius, or Hair-Worm. It is called by the last name because many persons, ignorant of its history, have supposed that it is a horse-hair which has been transformed into a worm!

MOLLUSKS.

The term Mollusk comes from a word which means *soft;* and these animals have a soft body with no backbone nor internal skeleton; nor is the body divided into rings or joints, as in the Articulates. Most of them have a hard covering called a shell, and are often called *Shell-Fish;* but they are in no way related to Fishes. The shells are the parts which we oftenest see; for when the animal is dead, the soft parts soon disappear, and only the shell remains. Curious and wonderful as the shells are, they often give only the faintest idea of the appearance of the animals when alive. See the differences between Figures 356 and

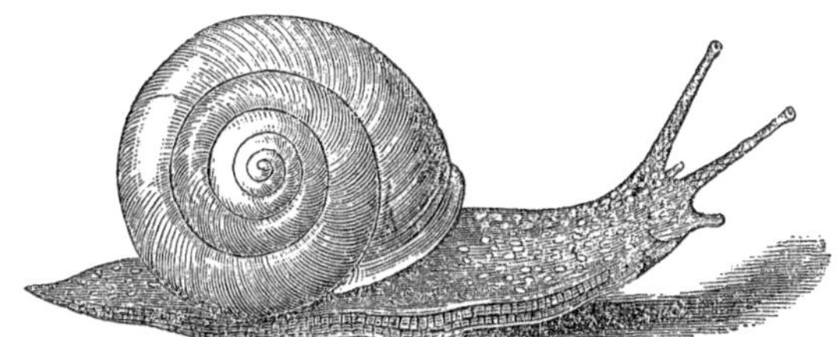

Fig. 356. — Helix, — alive.

357, where the first represents the shell alive and the animal expanded, the second the shell as when dead, or when the animal is concealed in the shell. It is important to know that the shell is a part of the animal, and not a mere house which it enters and leaves at pleasure; although it readily expands much beyond the limits of the shell, and withdraws itself wholly within the same

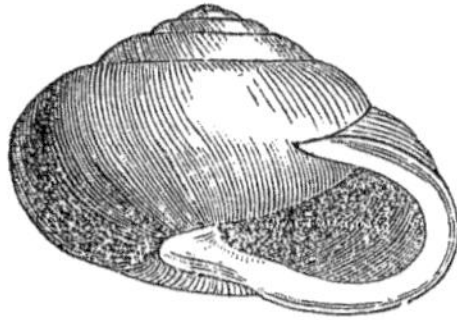

Fig. 357. — Helix, — dead.

again. Mollusks have, in a greater or less degree, the senses of the higher animals, though they greatly differ among themselves in this respect.

The kinds of mollusk are very numerous, — not less than fifteen or twenty thousand. They abound in the sea, on the marshes, in pools, streams, ponds, and lakes, and on the land; and they are full of interest when we study them, and all serve some important purpose. They are the food of many other animals. The Right Whale feeds upon small kinds which swim freely in the open sea; the Cod and Haddock and many other useful fishes fatten upon those gathered near or on the bottom; and sea-birds feast upon those left bare by the tide. Man reckons the Oyster, Clam, and Scallop among his choicest dishes; and in seasons of scarcity the poor inhabitants on many a sea-coast depend upon mollusks for a large part of their daily food. These animals also furnish the bait for all the extensive fisheries of the North Atlantic. Some of them yield rich dyes. The celebrated Tyrian purple of the ancients was obtained from shell-fish.

The shells of mollusks are limestone, or carbonate of lime. Pearly within, and even without when polished, and of soft and delicate colors, they are often exceedingly beautiful, and are eagerly sought for. The child gathers them for toys, and thinks he hears the roaring of the sea as he puts them to his ear; the savage wears them as ornaments, and some of them as marks of chieftainship; some kinds are gathered by civilized nations and used instead of money in trading with barbarous tribes; other kinds are gathered and wrought by skilful hands into almost numberless articles of use and luxury; and the true naturalist, more enthusiastic than all others, traverses sea and land, and cheerfully endures hunger, thirst, and fatigue, that his collection of shells may lack neither "Argonaut" nor "Nautilus," "Cone," "Cowry," nor "Wentle-trap," "Helix" nor "Limnæid," "Pecten," "Mother-of-Pearl," nor "Unio," nor any other which will enable him to understand more clearly this department of the animal kingdom, and the works of God as revealed in these wonderful objects.

ARGONAUTS, CUTTLE-FISHES, SQUIDS, AND NAUTILI, OR CEPHALOPODS.

These animals all live in the ocean, have a mouth armed with a stout beak, resembling that of a parrot, a large eye on each side of the head, and surrounding the mouth are long, muscular arms, or tentacles, covered with cup-like suckers, by means of which they cling with the greatest firmness to whatever they lay hold of, — it being easier to tear away an arm than to release it from its hold. They have within the body a sac containing an ink-like fluid, with which they cloud the water, and thus conceal themselves whenever they wish to escape from an enemy. The word Cephalopod means *head-footed*, and is given to these mollusks because their locomotive organs are attached to the head, as just described. Cephalopods vary from a few inches to several feet in length, according to the kinds. They have a most wonderful power of changing their colors, — their hues varying almost every moment. They swim by means of their arms, or with them crawl on the bottom with the head downwards. They are very voracious, eagerly devouring fishes and other animals, whose flesh they readily tear in pieces by their stout hooked beaks.

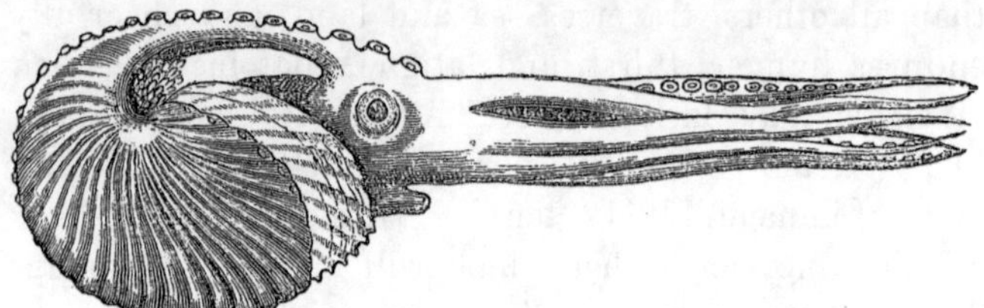

Fig. 358. — Argonaut, or Paper-Sailor. Much reduced.
Warm Seas.

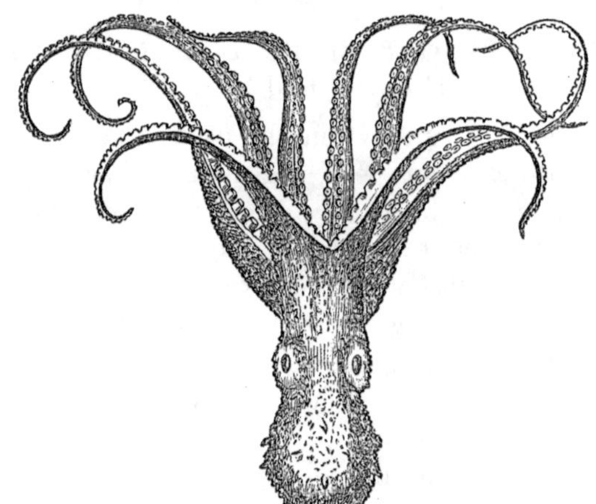

Fig. 359. — Octopus, or Poulpe. Much reduced.
Mediterranean.

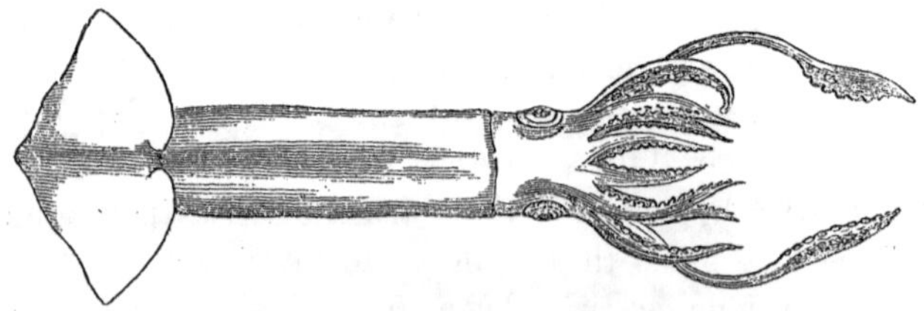

Fig. 360. — Squid or Loligo. Much reduced.
Atlantic Coast of United States.

If all accounts of them are true, cephalopods sometimes reach an enormous size. Aristotle tells us of one which was five fathoms in length! In 1853 a cuttle-fish, whose tentacles were five or six inches in diameter, was cast upon the shores of Jutland. In 1861 the officers and crew of the French steamer Alecton saw one, forty leagues northeast of Teneriffe, which was estimated to be at least fifteen feet in length, with arms five or six feet long, and a beak a foot across.

PAPER-SAILORS.

The Argonauts, or Paper-Sailors, Figure 358, have a very delicate and beautiful shell, and they swim by placing two of their arms, which are webbed, close to the sides of the shell, and the others close together, and then ejecting water from the funnel seen just below the eye. The Argonaut is often called Nautilus, — the true Nautilus is another animal, — and it has frequently been stated that it sails on the sea by spreading its sail-shaped arms to the breeze; a pleasant story, but one which naturalists no longer believe.

OCTOPUS.

The Octopus, or Poulpe, Figure 359, has no outside shell, and the arms are united at the base by a web. It varies from one or two inches to two feet in length, and has only eight arms.

SQUIDS, OR LOLIGOS, AND CUTTLE-FISHES.

Squids have a long body, and broad, fin-like organs at the hind extremity, and they have a long and slender internal shell which, from its shape, is called a "pen." They are from one to two feet and a half long, and, like cuttle-fishes, have ten arms, two of which are longer than the others. By filling their body with

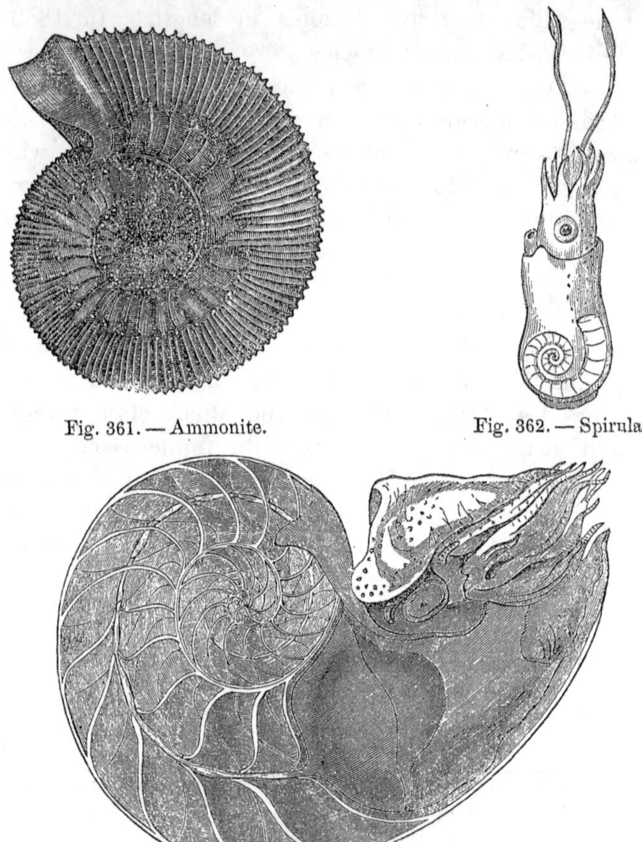

Fig. 361. — Ammonite.

Fig. 362. — Spirula.

Fig. 363. — Pearly Nautilus. Much reduced.
Pacific and Indian Oceans.

water, and then forcibly ejecting it, they send themselves backwards through the water with the swiftness of an arrow. Immense numbers are used for bait in the cod-fishery.

Cuttle-Fishes resemble Squids, but have two of the arms or tentacles much lengthened and expanded at their tips; and they have a broad, internal shell, called cuttle-bone. This is the "cuttle-bone" which is given to canary-birds. On the coasts of the Eastern Mediterranean cuttle-fishes are so abundant that the cuttle-bones are thrown up by the waves into ridges miles in length. Like other cephalopods, cuttle-fishes have the power of clouding the water by ejecting an inky fluid into it when they wish to escape. This ink, when dried and prepared, is the "sepia" used in painting.

SPIRULAS.

The Spirulas resemble those just described, but have a coiled shell inside, Figure 362, and the shell is divided by partitions into chambers.

NAUTILI AND AMMONITES.

The Nautilus is the only living Cephalopod which has an external chambered shell. Figure 363 shows the Nautilus as it appears when cut open; the animal is in the outer chamber, which communicates with all the others by means of a tube called the siphuncle. The animal has occupied each chamber in turn, moving forward, and making a partition behind as often as it outgrew its old home.

Ammonites, Figure 361, are chambered-shelled Cephalopods that lived in the seas ages ago; hundreds of kinds of these, from an inch to a yard in diameter, are found imbedded in the rocks of this and other countries.

SNAILS, OR GASTEROPODS.

The term Gasteropod means *stomach-footed*, and is given to these animals because the lower side serves them as a sort of foot, by means of which they creep along. But this "foot" is in no way related to the feet of the backboned animals. Most of the Gasteropods have a shell; and as this is made of only one piece, or valve, they are often called Univalves. Some, however, have no shell in the adult state, though all have a shell when first hatched. Most Gasteropods have a lid or door, called the *operculum*, with which they close the opening to the shell when they withdraw within. It is a horny plate, sometimes strengthened by shelly matter. Their eyes are two, and generally on long stalks, as seen in Figure 356; they perceive light,

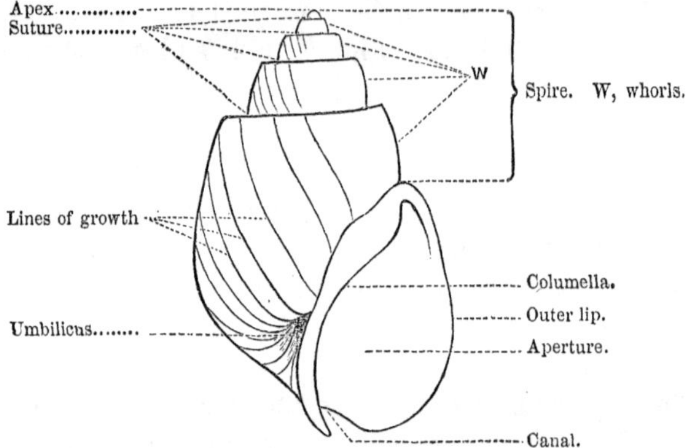

Fig. 364.—Names of the parts of a Gasteropod Shell.

but probably cannot distinguish objects. Many of the Gasteropods have horny jaws; but one of the most curious parts of these animals is the tongue, or lingual ribbon, which is a band armed with a great number of glossy silicious teeth, which are arranged in rows in the most regular manner, and differently in different kinds. The tongue of some kinds contains one hundred and sixty rows of teeth, and one hundred and eighty teeth in each row, or more than twenty-eight thousand in all!

Many of the Gasteropods feed upon vegetable substances, and these have the aperture of the shell entire. The others feed upon animal substances, and have the aperture notched, or drawn out into a canal, as in Figures 365–377. Some of these feed upon dead animals which they find; others attack living mollusks; and though the latter are shut tightly within their shells, the hungry Gasteropod, with its rasp-like tongue, files a neat round hole through the shell, and then leisurely feasts upon its contents. Thus clams and other large mollusks fall a prey even to some of the very small carnivorous gasteropods.

The Gasteropods are divided into Air-breathers or Pulmonifers, as Land-Snails, and the Water-breathers or Branchifers, as the Sea-Snails and River-Snails. The first look like the parents, only smaller, as soon as they are born; the young of the latter differ from their parents, and, instead of creeping, swim with a pair of fins springing from the sides of the head.

STROMBS, CONCHS, OR WING-SHELLS, &c.

These are large marine shells, some of them the largest of the Gasteropods. One kind, called the Fountain Shell, is extensively used for making shell-cameos; three hundred thousand of this kind were carried from the West Indies to Liverpool in a single year. The interior of the conch is of the richest rosy hue.

MUREX SHELLS.

Murex and its relatives are marine, and prey upon other mollusks. The Spiny Murex of the Moluccas, the Pyrula and Tritonium of the coast of the United

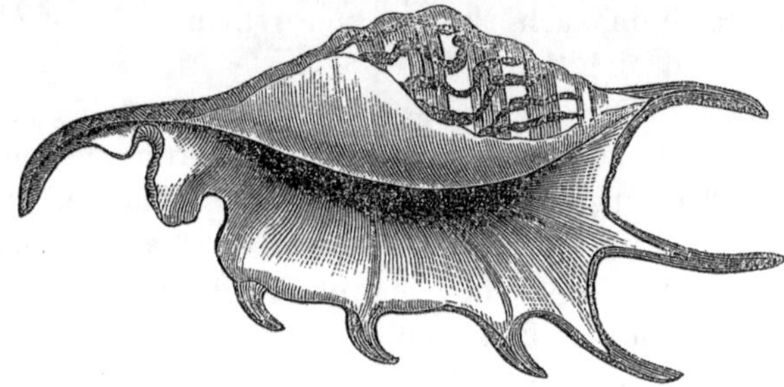

Fig. 365. — Scorpion Shell, or Pteroceras. Much reduced. Chinese Seas.

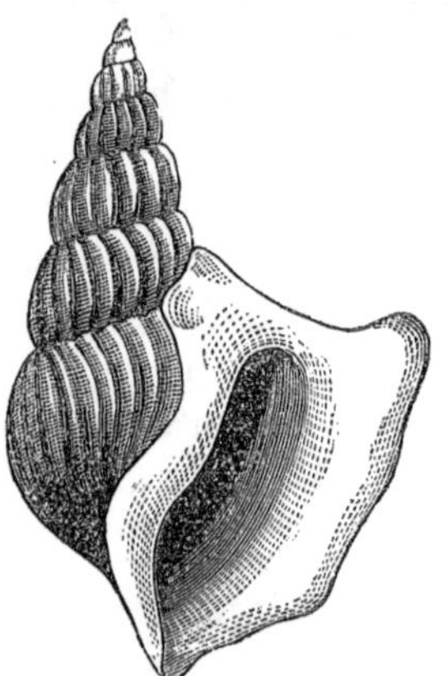

Fig. 366. — Aporrhais. Coast of New England.

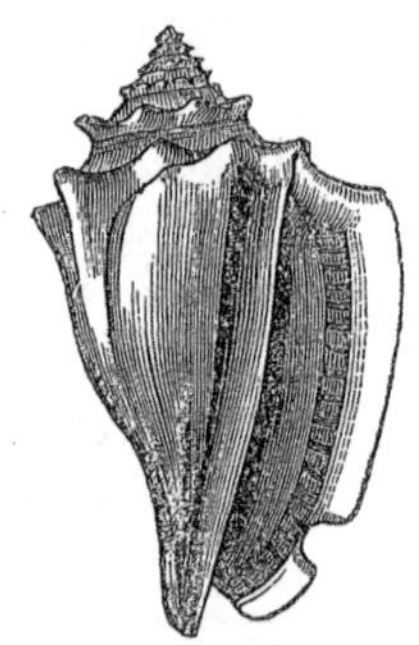

Fig. 367. — Stromb, or Conch. Much reduced. West Indies.

States, and the Frog Shell of Australia, are some of the principal ones. The ancients obtained the Tyrian purple dye from the Murex gasteropods.

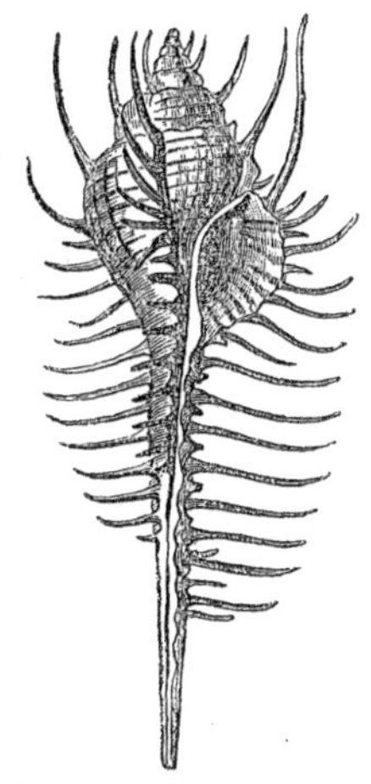

Fig. 368.—Murex. Much reduced. Molucca.

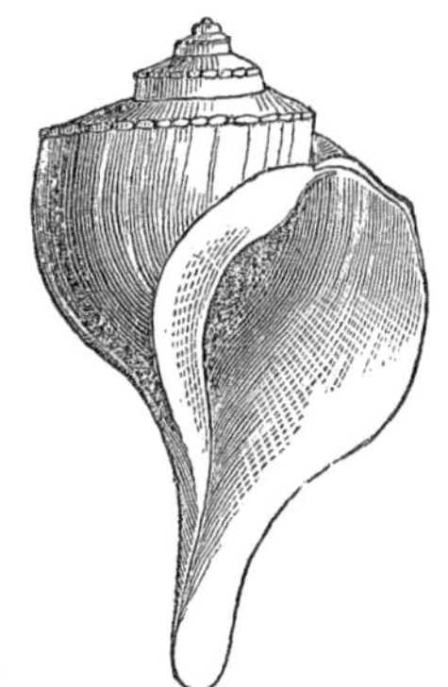

Fig. 369.—Pyrula. Much reduced. Coast of United States.

Fig. 370.—Tritonium. Coast of New England.

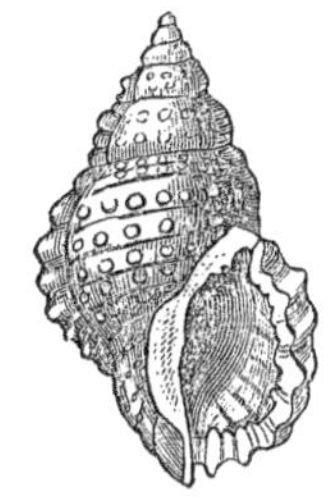

Fig. 371.—Frog Shell, or Ranella. Reduced. Australia.

WHELKS.

The Whelk is one of the most common of the Gasteropods. Figure 372 shows one species as it appears when crawling up the glass sides of the aquarium with the foot towards you. The Fusus, of the coast of the United States, may be found upon the shore after storms. The Harp Shell, of the Pacific, is always

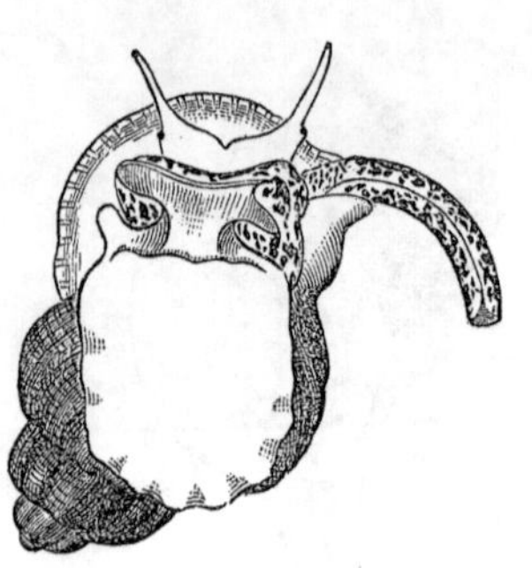

Fig. 372. — Whelk, or Buccinum. North Atlantic.

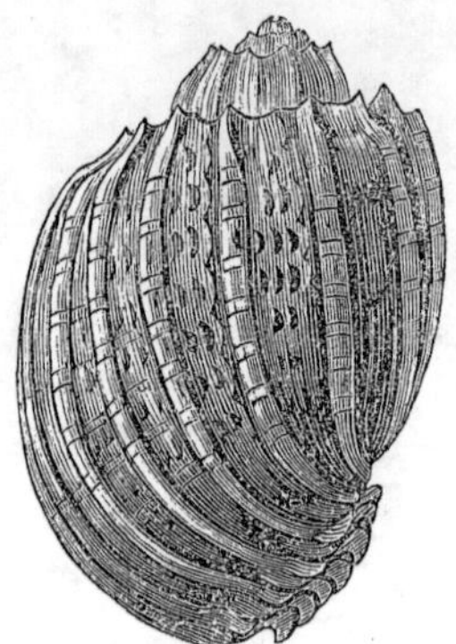

Fig. 373. — Harp Shell. Reduced. Mauritius.

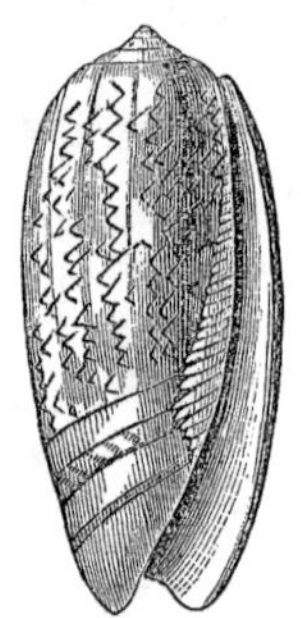

Fig. 374. — Oliva. Reduced. Panama.

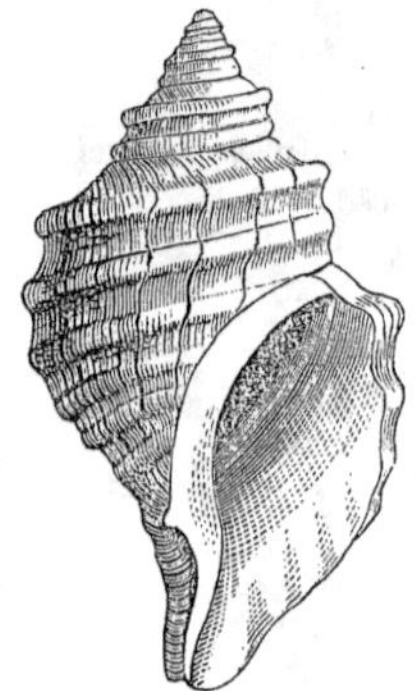

Fig. 375. — Fusus. United States.

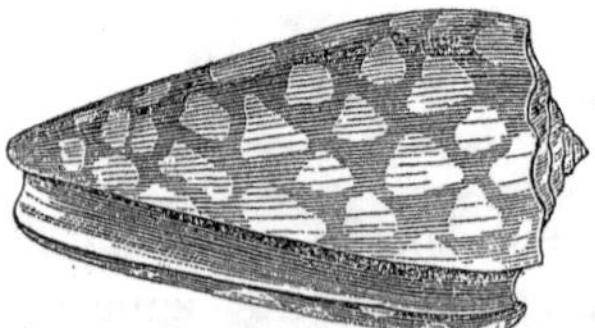

Fig. 376. — Cone Shell. Reduced. China.

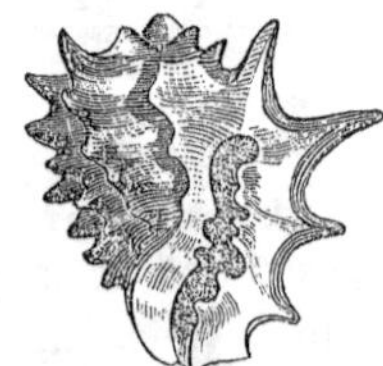

Fig. 377. — Ricinula.

admired for its beautiful form and its delicate colors. The Olive Shell, of Panama, is very beautiful, and is taken alive by bait attached to lines.

CONES.

There are nearly a thousand kinds of these Gasteropods, which are shaped like a cone with the top downwards.

VOLUTES.

The Volutes, Mitre-Shell, and Marginella belong under this head. Figures 378–380.

COWRIES.

The Cowries are abundant in the warm seas, and are found on reefs and under rocks. The shell has a shining enamelled surface, and many kinds are beautifully spotted and clouded. The Asiatic islanders use them to adorn their clothing, and for sinkers to their fishing-nets, and in trading. One kind, called the Money-Cowry, is brought in immense quantities from the Pacific to England, and then carried to Western Africa, where it is used for money in trading with the natives. This is a small kind scarcely an inch long. The Egg-Cowry and the Cypræa of the Indian Ocean show the general form of these shells. Figures 381–383.

NATICAS, PYRAMID-SHELLS, CERITHIUMS, &c.

The Naticas are sea-snails which have the shell somewhat globe-shaped. The Pyramid-Shells are so named from their shape. The Cerithiums are named from a word which means a *horn*. The Melanias are fresh-water shells, common in the Western and Southern States.

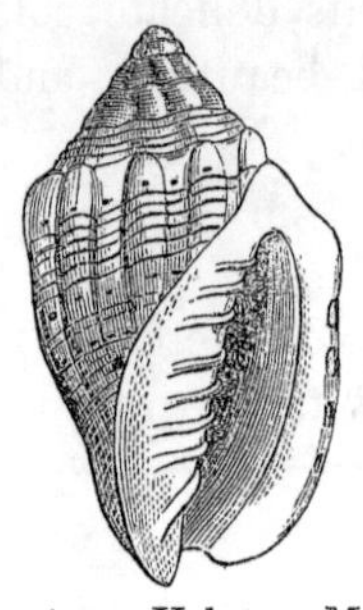

Fig. 378.—Volute. Much reduced. West Indies.

Fig. 379.—Marginella. Reduced. W. Africa.

Fig. 380.—Mitre Shell. Much reduced. Ceylon.

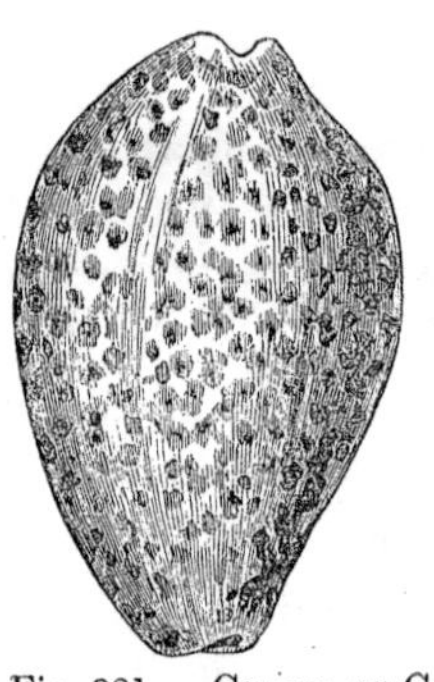

Fig. 381.—Cowry, or Cypræa. Much reduced. Indian Ocean.

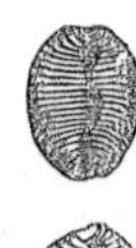

Fig. 382.—Trivia. Britain.

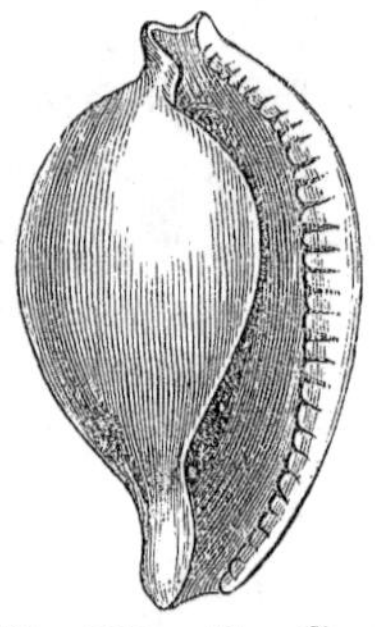

Fig. 383.—Egg Cowry. Much reduced. New Guinea.

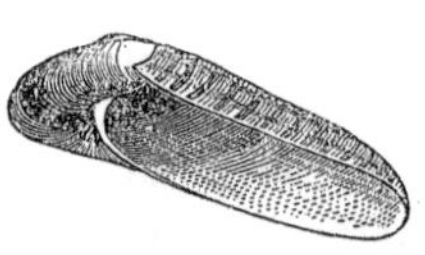

Fig. 384.—Sigaretus. West Indies.

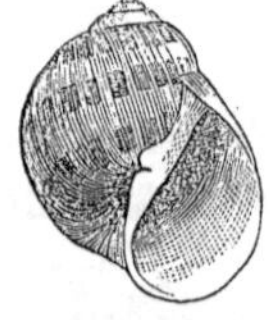

Fig. 385.—Natica. Coast of N. England.

Fig. 386.—Pyramid-Shell. Reduced. Britain.

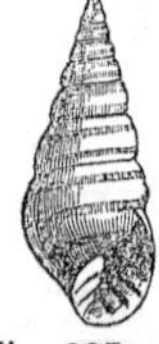

Fig. 387.—Pyramid-Shell. W. Indies.

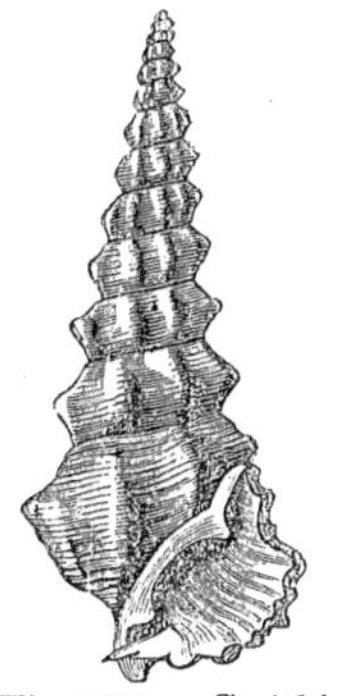

Fig. 388.—Cerithium. Much reduced. Molucca.

Fig. 389.—Melania. Western States.

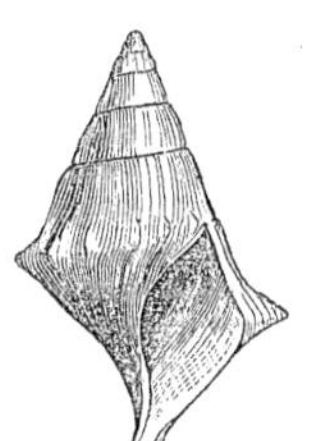

Fig. 390.—Io. Southern States.

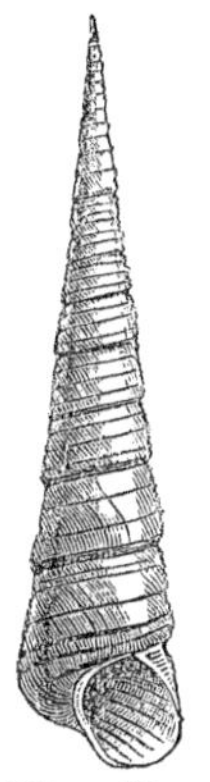

Fig. 391.—Tower-Shell, or Turritella. West Indies.

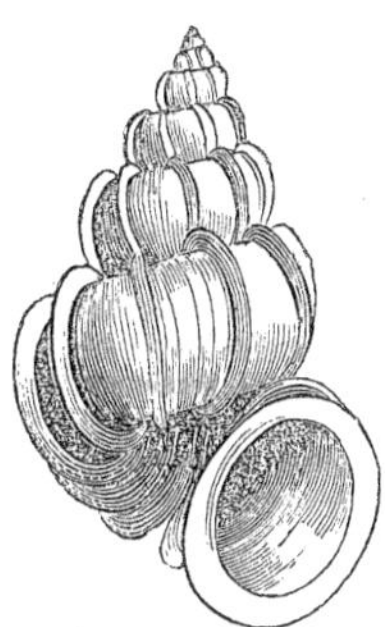

Fig. 392.—Wentle-trap. Reduced. China.

Fig. 393.—Worm-Shell, or Vermetus. West Indies.

Fig. 394.—Periwinkle, or Litorina.

Fig. 395.—Lacuna.

Fig. 396.—Valvata. U. States.

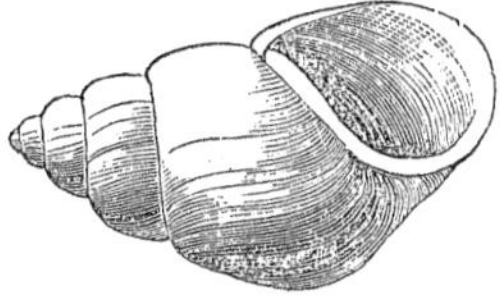

Fig. 397.—River Snail-Shell, or Paludina. United States.

WENTLE-TRAPS, &c.

The Tower-Shell and the Worm-Shell of the West Indies, and the true Wentle-traps of the tropical and temperate seas, belong in this group. The Royal Staircase, or Wentle-trap, Figure 392, was formerly very valuable. A specimen has been sold for a hundred pounds sterling, although it is now worth only a few dollars.

PERIWINKLES.

Periwinkles live in the sea near the shore. Two species are shown in Figures 394 and 395. They feed on algæ, — marine plants.

RIVER SNAILS.

These live in fresh waters, and have the shell covered with a green skin. They bring forth their young alive, and the little embryo snails, even before they are born, and when so small that they can scarcely be seen without a microscope, have a perfectly formed shell, a "foot" and operculum, delicate tentacles, and distinct black eyes.

VIOLET-SNAILS.

The Violet-Snails live together in large numbers, in the open sea, where they float by means of many air-vessels, which form a raft, *a*, Figure 404. The shell is thin, the base deep violet color, and the spire almost white. They yield a violet dye.

LIMPETS.

Limpets are found clinging tightly to stones and other shells, and move about but little or not at all.

They are all marine. On the coast of England the Limpet is much used by fishermen for bait, and on the coast of Berwickshire twelve millions have been collected yearly for this purpose. In the north of Ireland the people collect it for food. On the western coast of South America there is a kind of Limpet which is a foot across, and the natives use its shell for a basin.

LAND-SNAILS.

Land-Snails are very numerous, more than four thousand kinds being already known. Figures 409–414. They all feed upon decaying plants. One of the largest and one of the most common is the *Helix albolabris*, Figure 409. It is easily found by searching under old logs, stumps, and leaves. In warm, damp weather, snails of this and similar kinds come out of their hiding-places, and may be seen crawling over the leaves and up the trunks of trees. In early summer they lay their eggs in the loose soil beside logs or stones, and in twenty or thirty days the young hatch. When the cold weather of autumn comes they seek a sheltered spot, close the mouth of the shell with a thin membrane which they secrete, and at length become torpid, and remain in that condition till the warm days of the following spring.

POND-SNAILS, OR LIMNÆIDS.

These live in fresh waters, and lay their eggs in transparent masses on aquatic plants and on stones. They have a thin and horn-like shell. Figures 415–417. They feed on plants, and glide along the surface of the water shell downwards. They thrive well in an aquarium, where they are also very useful, as

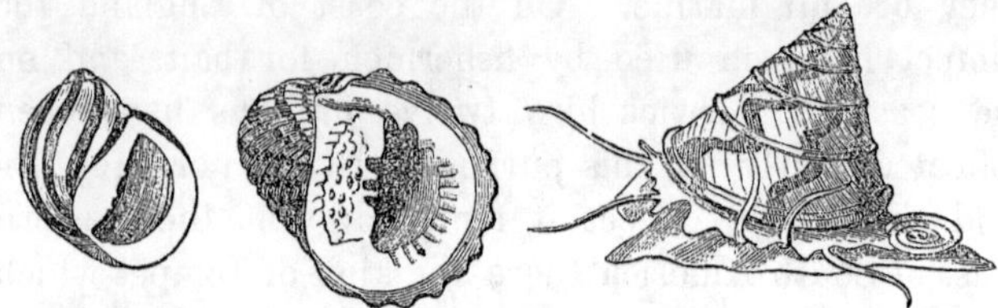

Fig. 398. — Neritina. Pacific.

Fig. 399. — Nerita. Scinde.

Fig. 400. — Trochus. Britain.

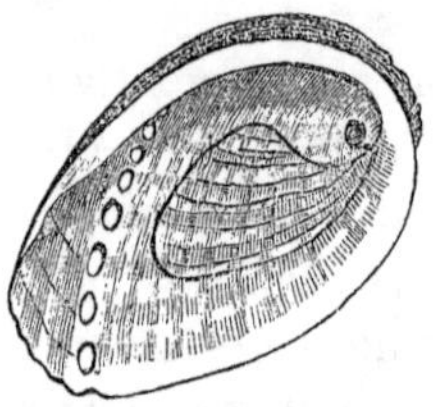

Fig. 401. — Ear-Shell, or Haliotis. Reduced. Britain.

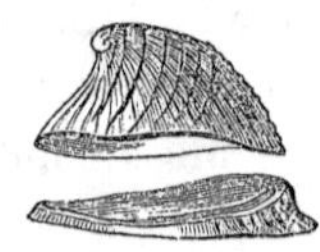

Fig. 402. — Cup-and-Saucer Limpet. Philippines.

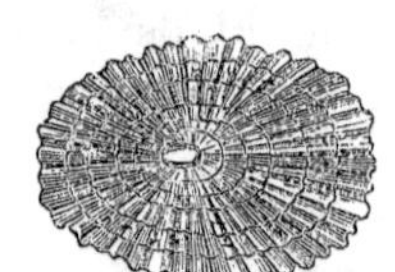

Fig. 403. — Key-hole Limpet. West Indies.

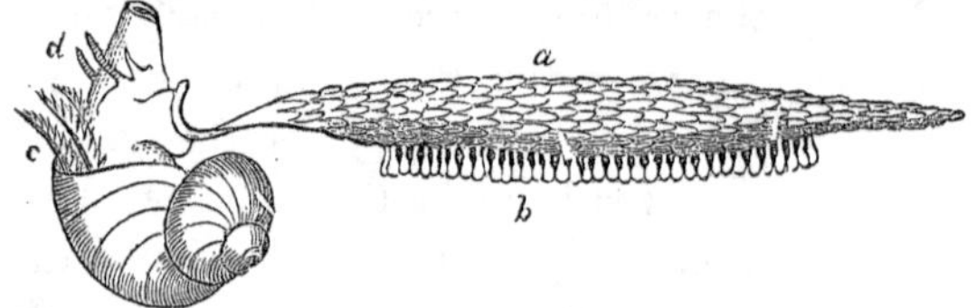

Fig. 404. — Violet-Snail. Atlantic.
a, raft; *b*, egg capsules; *c*, gills; *d*, tentacles and eye-stalks.

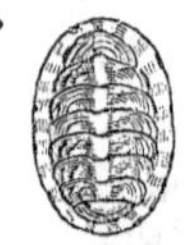

Fig. 405. — Chiton. Coast of N. England.

Fig. 406. — Rock Limpet, or Patella. Coast of New England.

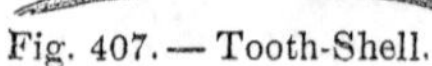

Fig. 407. — Tooth-Shell.

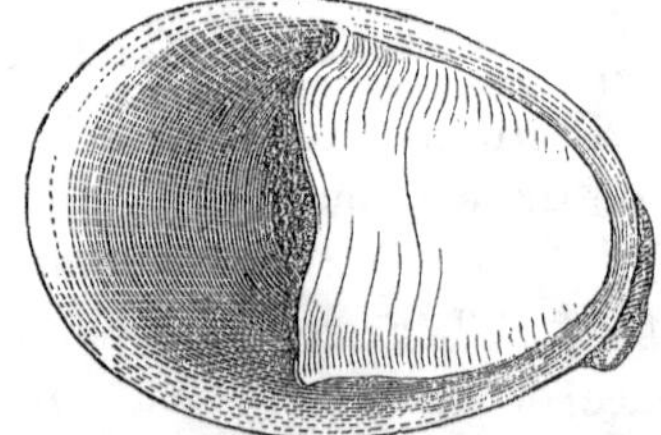

Fig. 408. — Crepidula. New England

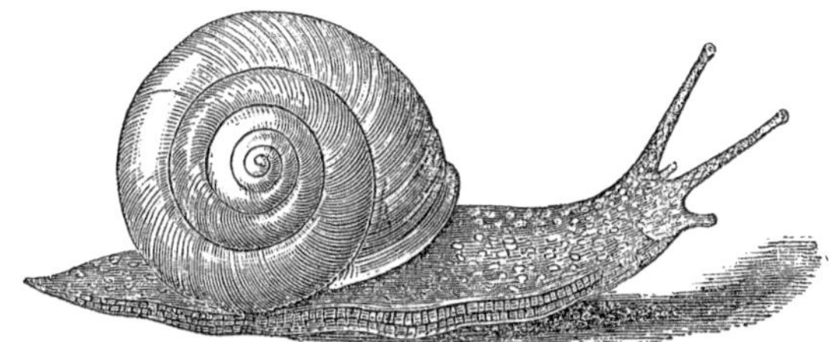

Fig. 409.—Helix.

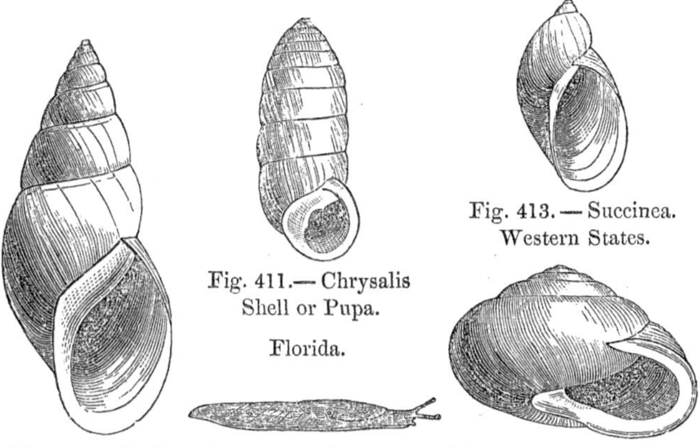

Fig. 410.—Bulimus. California.

Fig. 411.—Chrysalis Shell or Pupa. Florida.

Fig. 412.—Slug, or Limax. N. England.

Fig. 413.—Succinea. Western States.

Fig. 414.—Helix. Northern States.

Fig. 415.—Physa. United States.

Fig. 416.—Planorbis. United States.

Fig. 417.—Limnæa. United States

they eagerly devour the green confervæ that grow on the sides of the glass.

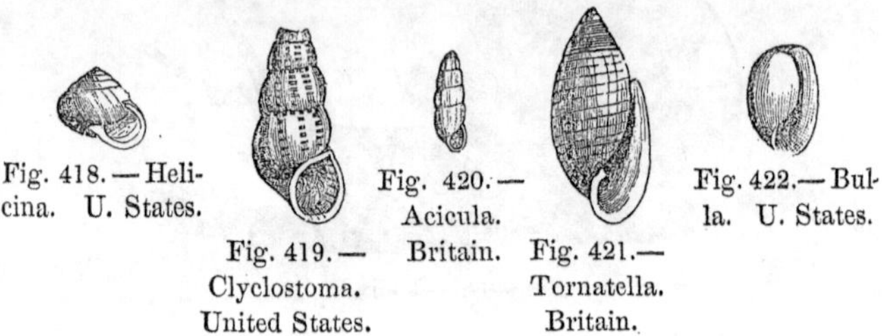

Fig. 418.—Helicina. U. States.

Fig. 419.—Cyclostoma. United States.

Fig. 420.—Acicula. Britain.

Fig. 421.—Tornatella. Britain.

Fig. 422.—Bulla. U. States.

SEA-SLUGS.

These have no shells, and many of them only slightly resemble the Gasteropods before described. See Figures 423–426.

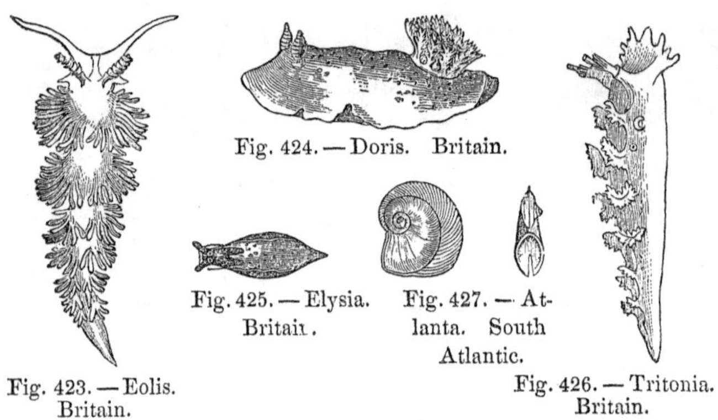

Fig. 423.—Eolis. Britain.

Fig. 424.—Doris. Britain.

Fig. 425.—Elysia. Britain.

Fig. 427.—Atlanta. South Atlantic.

Fig. 426.—Tritonia. Britain.

HETEROPODS AND PTEROPODS.

These live in the open sea. Some of them move in immense swarms, miles in extent. Figures 428–430. They much resemble the young of ordinary sea-snails. They form the principal food of the Right Whale. One kind, the Clio, Figure 430, is said to have upon the head three hundred and sixty thousand suckers!

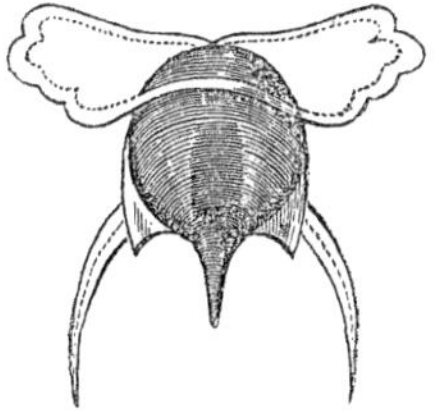
Fig. 428. — Hylea. Atlantic.

Fig. 429. — Limacina. South Polar Seas.

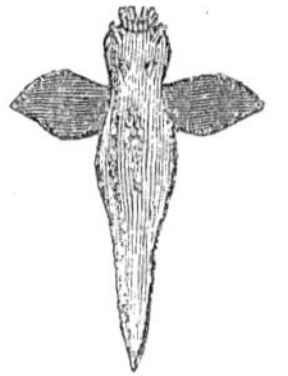
Fig. 430. — Clio. Arctic.

ACEPHALS, OR HEADLESS MOLLUSKS.

These mollusks seem to have no head, this part of the body being concealed within, and only faintly shown, as in Clams, Oysters, Mussels, &c.

BIVALVES.

These are acephals which have a shell composed of

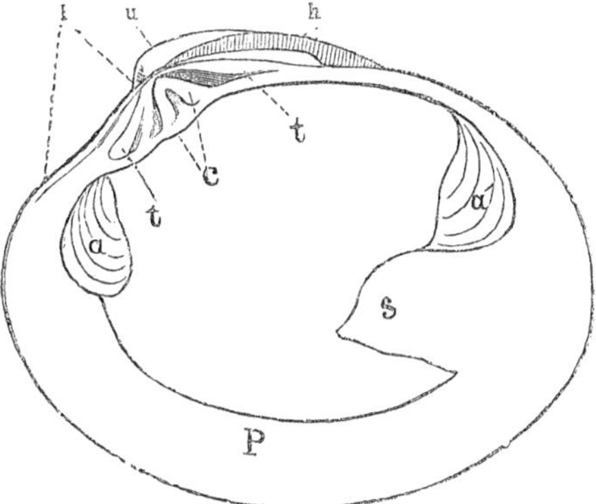

Fig. 431. — One valve of a Bivalve Shell, showing the names of the parts.

a, anterior retractor muscle ; *á*, posterior retractor ; *t*, lateral teeth ; *c*, cardinal tooth ; *l*, lunale ; *u*, umbo ; *h*, hinge ligament ; *s*, retractor of siphons ; *p*, pallial impression.

two pieces, or valves, joined together on one side by a hinge, and held tightly together by one or two strong muscles which pass from one valve to the other on the inside. When the animal relaxes these muscles the shell is forced open by an elastic body called a ligament, situated at the hinge. Some kinds live in the sea, others in brooks, rivers, ponds, and lakes. Some idea of them all may be gained by studying the common mussel, Figure 437, of the brooks, or the common clam, Figure 452, of the sea-coast. Take the clam: place it in a large basin of sea-water, and soon it will begin to put out a dark-colored organ as long as the shell, — it can stretch it out two or three times the length of the shell. This is supposed by most persons to be the head, but it is not; the head is within the shell and at the opposite end. At the end of the dark organ are two holes, — one larger than the other, — these being the openings of two tubes which are enclosed in the dark-colored sheath; and around each opening there is a row of fringes or tentacles. A current of water is all the time flowing into the larger opening, and another current flowing out of the smaller opening. The first carries in pure water to supply air to the gills, and minute plants and animals to supply the mouth and stomach with food, and the outgoing current bears away the impure water together with the waste particles which the animal throws off. The currents are caused by a vast number of hair-like fringes which cover the gills within the mollusk, and which are constantly in motion. The position and appearance of the siphonal tubes in fresh-water mussels are seen in Figure 437.

Though mainly small, or of ordinary size, a few bivalves are very large. In the church of St. Sulpice, in Paris, the valves of a Tridacna weighing five hundred pounds, and two feet across, are used as vessels for the holy water. The Tridacna lives in the Pacific and Indian Oceans.

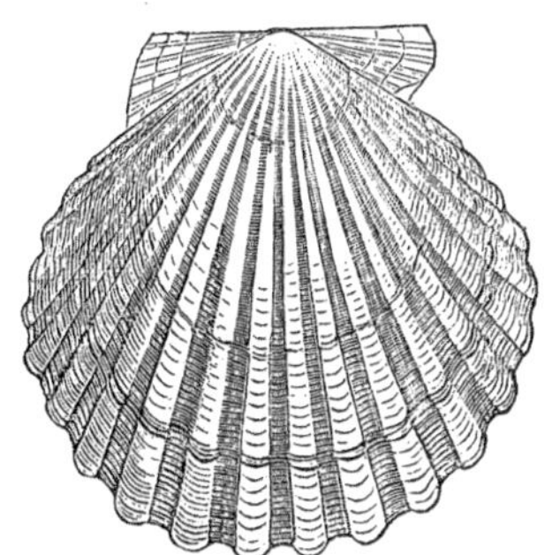

Fig. 432. — Pecten. From Cape Ann southward.

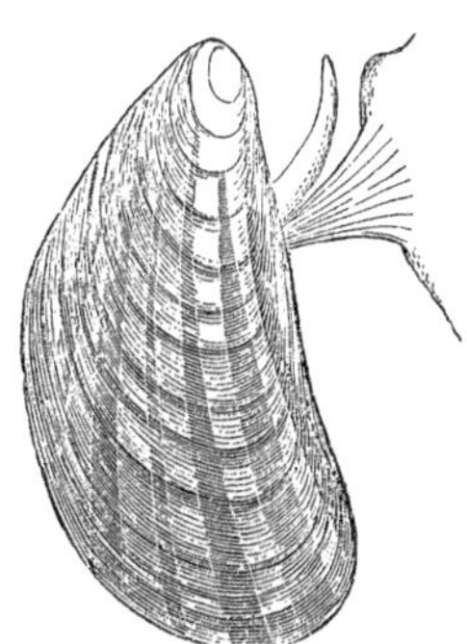

Fig. 433. — Mytilus. Both shores of the Atlantic.

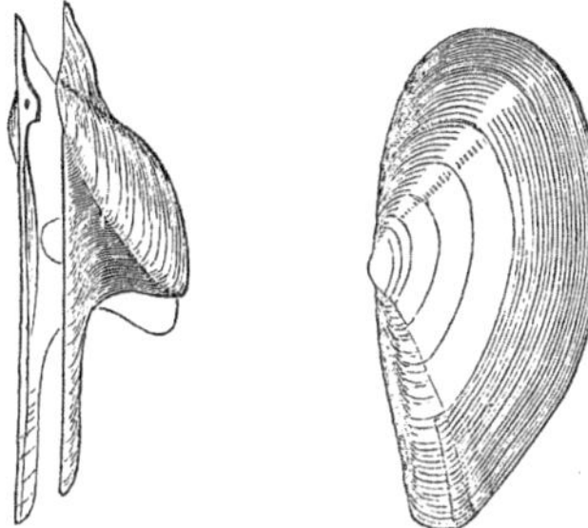

Fig. 434. — Avicula. Reduced. Mediterranean.

Fig. 435. — Leda. New England.

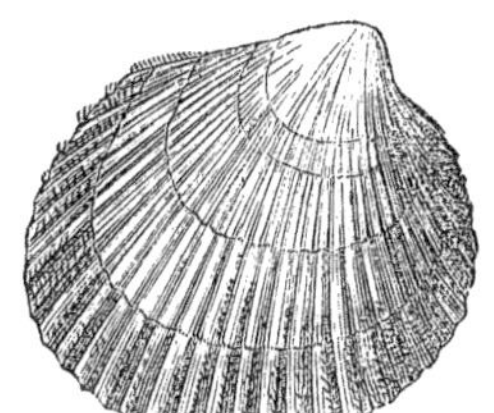

Fig. 436. — Cardicum. Reduced. New England.

OYSTERS, PECTENS, MUSSELS, PEARL-OYSTERS, &c.

Oysters are more highly prized for food than any other mollusks. They occur in the greatest quantities on the coast of the Middle States, especially in Delaware and Chesapeake Bay.

Pectens, or Scallops, Figure 432, are also prized for

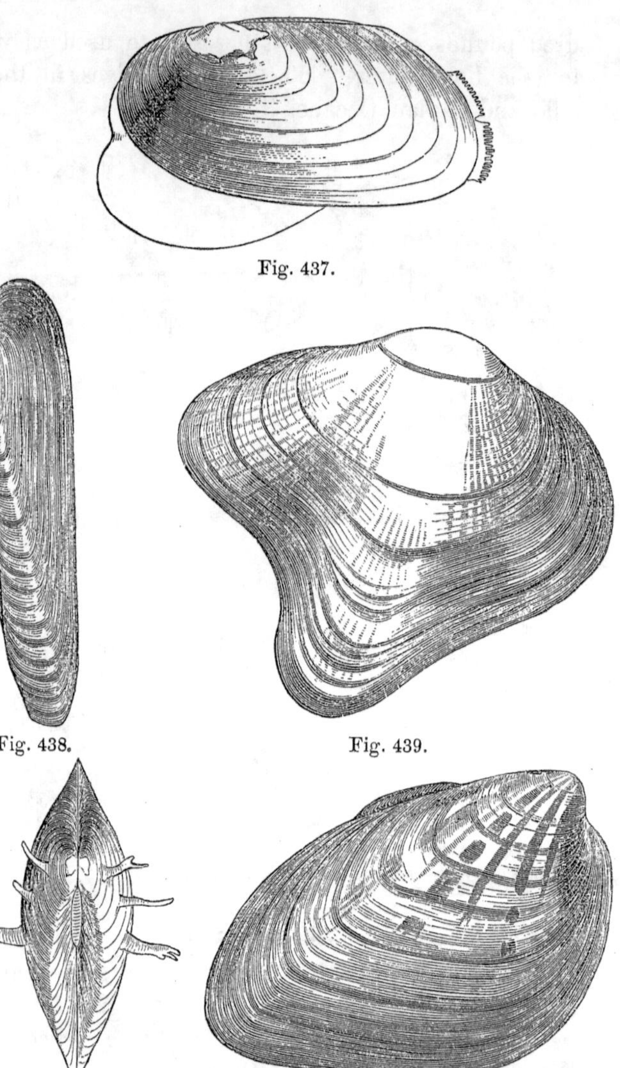

Fig. 437.

Fig. 438.

Fig. 439.

Fig. 440.

Fig. 441.

Figs. 437-441. — Unios. Reduced. United States.

food. Their beautiful shells are known to almost every one; for they are much used in making card-holders, pin-cushions, and other little articles both useful and pleasant to see. The pecten swims rapidly by opening and shutting its valves.

Sea-Mussels, Figure 433, inhabit mud-banks which are uncovered at low water. They multiply rapidly, and grow to their full size in one year. By means of a collection of horny threads, called a *byssus*, they attach themselves to rocks, or to the ground.

Pearl-Oysters, or Aviculas, Figure 434, have shells which make the beautiful material called "Mother-of-Pearl," which is extensively used for ornamenting fine cabinet-work, and for making knife-handles, paper-cutters, buttons, and a great number of other useful and beautiful articles. They also yield the Oriental pearls.

UNIOS, OR RIVER-MUSSELS.

These mollusks abound in brooks, rivers, ponds, and fresh-water lakes. They are sometimes called Naïdes, and there are very many kinds. It would take several books larger than this one to describe all the kinds found in the United States. A few of the forms of Unios are shown in Figures 437–441. Sometimes beautiful and valuable pearls are found in these mollusks. One of the pearls in the Royal Crown of England came from a river-mussel.

RAZOR-SHELLS, CLAMS, &c.

The Razor-Shells are very long and smooth. They burrow in the sand, and are good for food. The Common Clam burrows in sand and mud, and is extensively used for food, and for bait for cod.

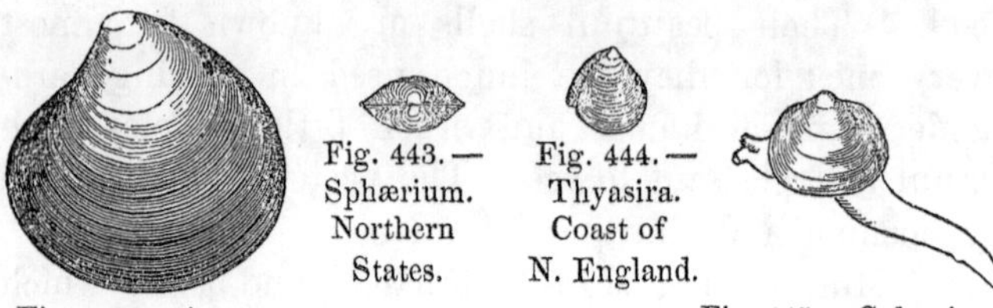

Fig. 442. — Astarte. Coast of New England.

Fig. 443. — Sphærium. Northern States.

Fig. 444. — Thyasira. Coast of N. England.

Fig. 445.— Sphærium. Northern States.

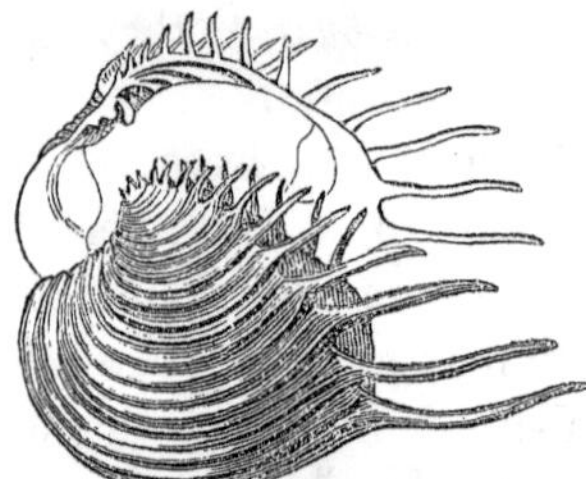

Fig. 446. — Cytherea. Reduced. West Indies.

Fig. 447. — Mactra. Britain.

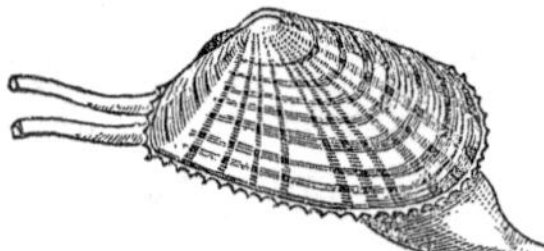

Fig. 448. — Tellina. Britain.

Fig. 449. — Tellina. Our coast.

Fig. 450. — Tellina. Our coast.

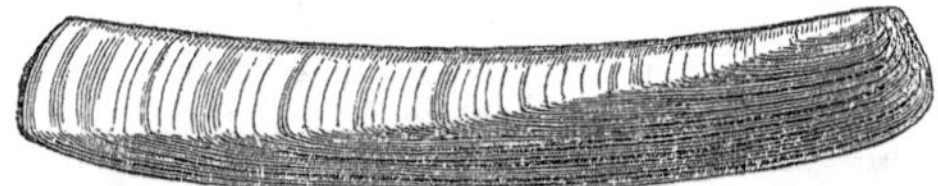

Fig. 451. — Razor-Shell, or Solen. Much reduced. Both shores of the Atlantic.

PHOLADS AND SHIP-WORMS.

Pholads have the shell very hard and rough, like a rasp, and they burrow in all sorts of substances, even in stone. Ship-Worms are long mollusks, looking like worms. The common kinds are about a foot long, but one kind is three feet in length. They bore into the timber of ships and wharves.

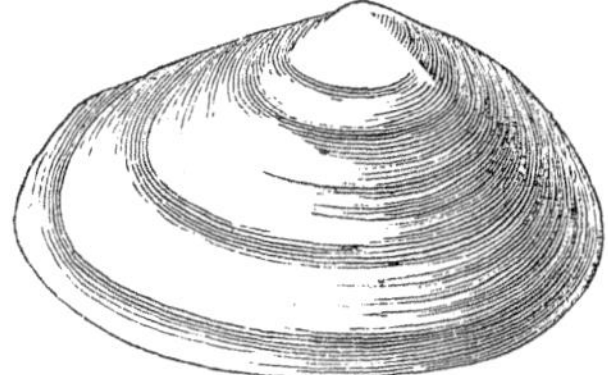

Fig. 452. — Common Clam. Reduced. Coast of New England.

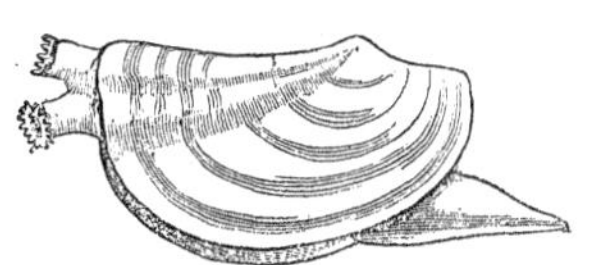

Fig. 453. — Pandora.

Fig. 454. — Gastrochæna. Galway.

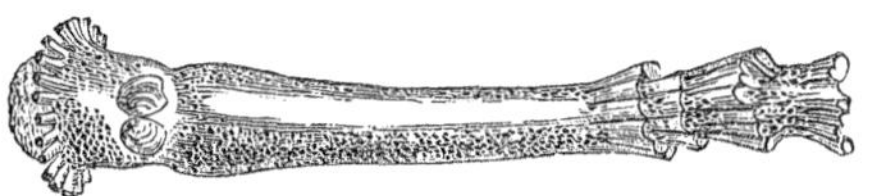

Fig. 455. — Watering-pot Shell. Much reduced.

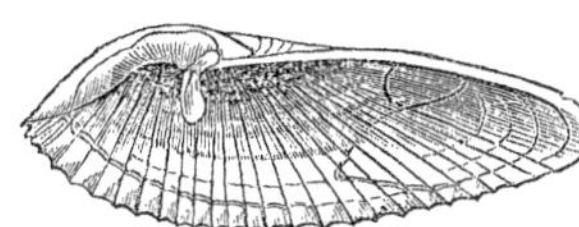

Fig. 456. — Pholas. Reduced. India.

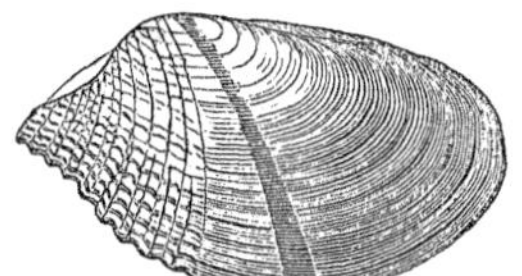

Fig. 457. — Pholas. New England and eastward.

TUNICATES.

Fig. 458. — Tunicate.

These are mollusks which have no shell, but are covered with a tough tunic, or skin. Sometimes they grow in clusters, attached by a stem to seaweed, rocks, or floating timber. They vary from the size of a pea to an inch or more in diameter. They are sometimes called Ascidians, from a word which means a *leather bag*.

BRACHIOPODS.

These mollusks have the two valves of unequal size, and in one of them there is a hole through which passes

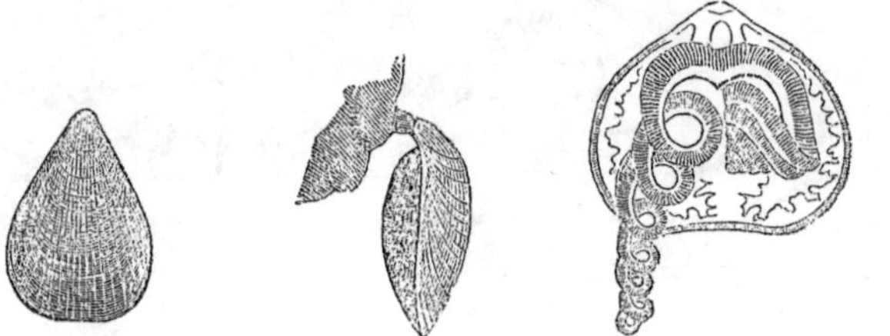

Figs. 459. — Terebratula, — a Brachiopod.

Fig. 460. — Brachiopod.

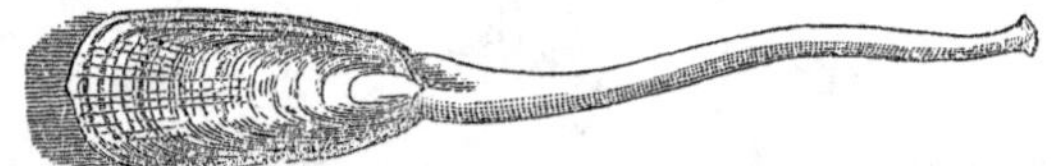

Fig. 461. — Lingula, — a Brachiopod. Reduced. Philippines.

a fleshy stalk, by which the shell is attached to the rocks. The word Brachiopod means *arm-footed*, and is given to these animals on account of the long, fringed arms growing from the sides of the mouth, and by means of which they make currents in the water and thus secure their food.

POLYZOANS.

These are very small or minute mollusks, growing in clusters upon shells, rocks, and other objects, both in the sea and in fresh waters, and which look very much like Polyps. They are often called Bryozoans.

Dr. Leidy and Captain Hyatt have described and beautifully figured many of our fresh-water kinds, and we hope you will some time see and read their interesting and instructive papers.

RADIATES.

THESE animals are so constructed that their parts radiate from a centre or central axis. In most of them the radiation is very plain. They all live in the water, and nearly all live in the sea, and are known as Echinoderms, Jelly-Fishes, and Polyps.

ECHINODERMS.

The word Echinoderm means *hedgehog-skin*, and is given to these animals because many of them have the outside covered with spines; thus reminding us of the hedgehog of the fields, which was described on page 45. It is an interesting fact that in the Radiates the parts are generally arranged according to what is called a reigning number; in Echinoderms this number is generally *five;* that is, the parts of each kind are five, or some multiple of five.

HOLOTHURIANS.

The Holothurians, or Sea-Cucumbers, have no spines,

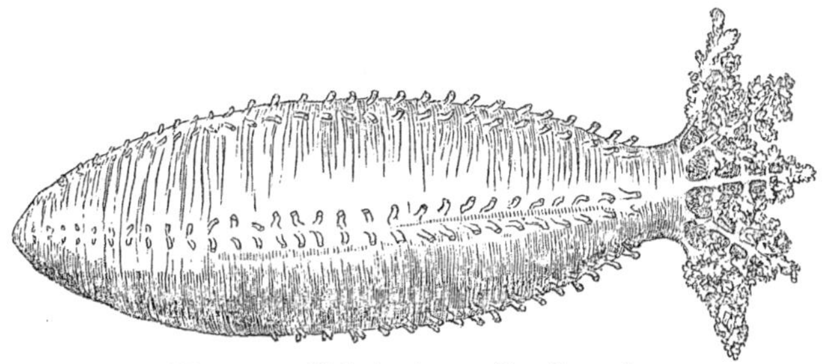

Fig. 462. — Holothurian, or Sea-Cucumber.

but are covered with a tough skin capable of great expansion and contraction, and containing particles of limestone. There are many kinds, varying from an inch to a foot in length. They live in the sea and are exceedingly interesting, and very beautiful when the long and delicate fringes around the mouth are expanded. When taken from the water they shrink and lose their beauty of form and color. They must be seen in the ocean, or in the aquarium, in order to get a good idea of them. Figure 462 shows one kind which is very common at Grand Menan, and Eastport, and other places in the North Atlantic. The Chinese call these animals *Trepang*, and use them for food.

SEA-URCHINS, OR ECHINOIDS.

True Sea-Urchins are hemispherical, or flattened, and have a hard shell composed of plates which are regular in form and firmly bound together. Upon

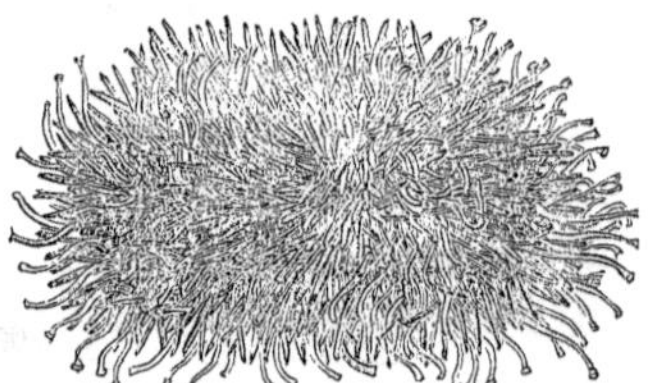

Fig. 463. — Sea-Urchin.

these plates are tubercles, and on these tubercles hard spines. In certain plates there are rows of holes through which pass fleshy organs called suckers, or ambulacra, with the end slightly expanded. By means of these suckers, which can be extended much beyond the spines, these animals can cling firmly to other bodies, and thus move about over the rocks, even up and down their smooth sides, as well as on level surfaces. So much can these suckers be extended that a Sea-Urchin has been seen to put them forth from the top, and, bending them downwards, cling to the bottom of the basin in which the animal was lying! Figure 463 shows a common kind of Sea-Urchin as it appears when alive. When the animal dies, the skin, which covers the shell and holds the spines in their

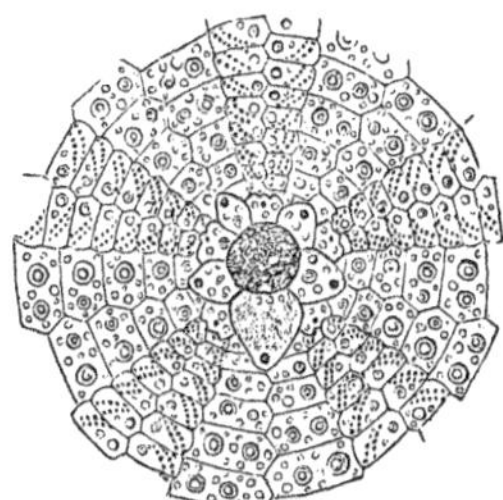

Fig. 464.—Top View of Sea-Urchin. Spines removed.

places, dries up, and the spines fall off, and then the shell, with all its beautiful structure and markings, is plainly seen. In the one represented in Figure 464 we find ten double rows of plates which run along the curved surface from the bottom to the top of the shell. In five of these double rows the plates are large, without holes, and are covered with large tubercles. Alternating with the double rows of large plates are five double rows of smaller ones, bearing few and small tubercles, and each plate is perforated with the holes for the suckers. The plates which bear the holes are called the *ambulacral* plates,—from a Latin word which means a *walk*, or *alley;* and the large plates without holes are called the *interambulacral* plates. At the termination of each of the five belts or zones of ambulacral plates there is a little triangular plate with a minute opening which marks the place of the eye. Alternating with these ocular plates, so called, are five larger plates, each being perforated with a larger hole through which the eggs are laid. One of these plates is much larger than the others, and is filled with very minute holes, and is called by naturalists the madreporic body. It is believed to serve as a filter or strainer to the water which passes through it into the body of the animal. The mouth, at the under side, is armed with five strong pointed and polished teeth, which form the outer part of a remarkable dental apparatus, which is called Aristotle's lantern. In a sea-urchin of ordinary size there are five or six hundred plates, all fitting together in the most perfect manner, and bearing more than four thousand spines; and the suckers number nearly two thousand!

Besides the spines and the suckers, there are scattered over the body and around the mouths of Sea-Urchins a great number of curious little organs called *Pedicillariæ.* They look like a stem ending in a knob, but the knob is composed of three pieces or blades, which open and shut tightly, thus forming a sort of pincers. The uses of these organs are not well understood.

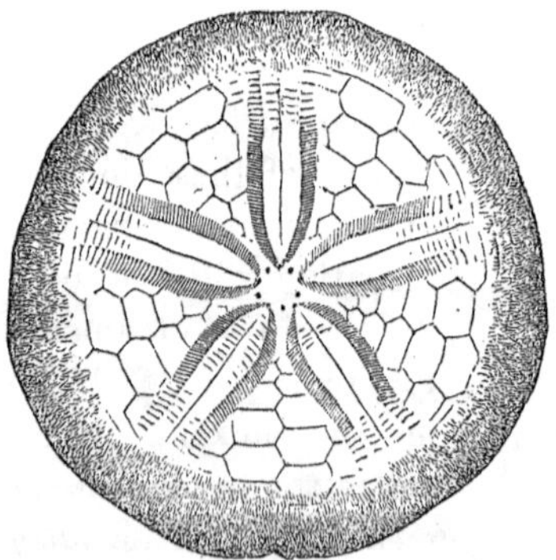

Fig. 465.—Echinarachnius. Northeast coast of North America.

The number of kinds of Sea-Urchins is quite large, and they vary in size from an inch to three or four inches

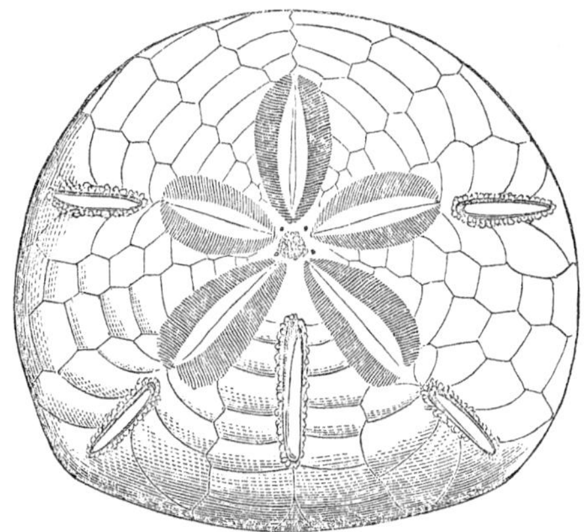

Fig. 466. — Mellita. Southeast coast of United States.

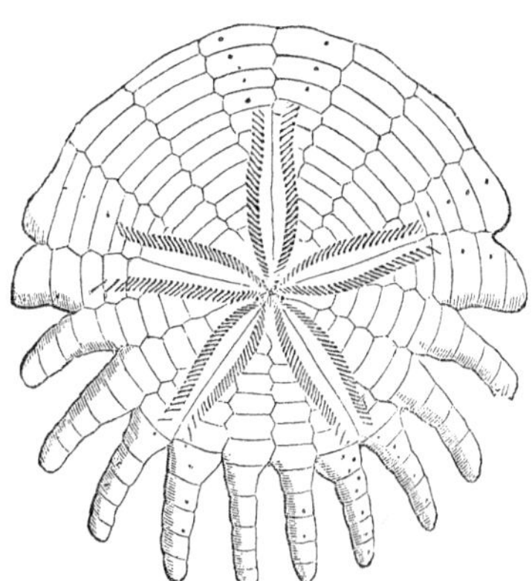

Fig. 467. — Rotula. Coast of Africa.

in diameter, and have spines from a quarter of an inch long to three or four inches in length. Some of them are capable of making holes in hard substances, even in limestone and granite.

Other kinds, like Figures 465, 466, 467, burrow in the sand. These are much flattened.

STAR-FISHES, OR SEA-STARS.

Star-Fishes are common on all rocky coasts. They are readily found by looking under the sea-weed in pools that have been left by the tide. They are so named from their star-like form, the disk or central portion gradually merging into the rays. Beneath each ray there is a large number of locomotive suckers, like those of the sea-urchins already described. These tubes are seen in Figure 468, where the upper

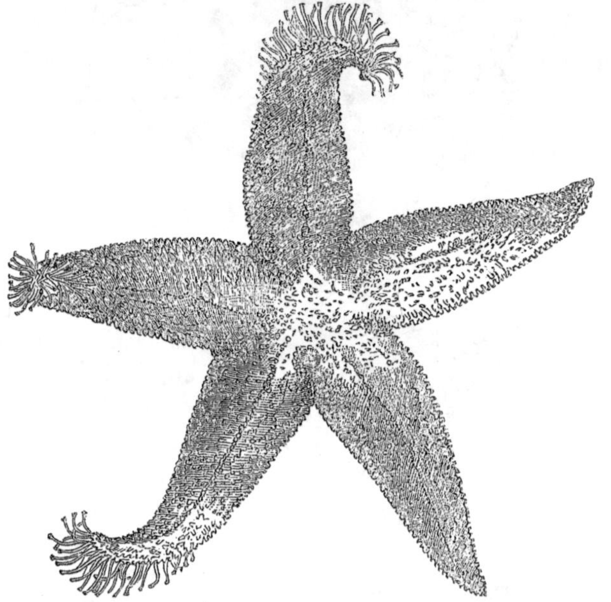

Fig. 468. — Star-Fish.

part of the Star-Fish is towards you, and three of the rays slightly turned backward. The mouth is on the under side in the centre, and there is an eye, or eye-spot, at the end of each ray. By means of the ambulacral tubes Star-Fishes move slowly but surely over the rocks and all kinds of surfaces, and they can cling to the rocks so firmly that they are often removed with difficulty, and will sometimes even allow their ambulacra to be pulled off rather than let go their hold. Their covering is not solid as in the Sea-Urchins, but is composed of movable plates, so that these animals are able to bend themselves in every direction, and thus work their way into holes and fissures in rocks where we should hardly expect to find them. Star-Fishes feed upon mollusks and other marine animals, and when they feed they turn the stomach out of the mouth and over the food to be devoured. A curious spot is seen on the back near the junction of two of the arms. This is the madreporic body described in speaking of the Sea-Urchins. It is a sort of minute sieve, and forms an entrance to a series of internal water-tubes, some of them connecting with the locomotive suckers and supplying them with water. Water is also admitted into the body through minute pores which cover the whole surface of the animal. Star-Fishes often lose one or more of their arms, or rays, by being dashed against the rocks by the waves, or the arm is bitten off by a fish. In all such cases a new one sprouts out in the place of the old one, and specimens may be found showing such new rays in all stages, from those that have just begun to sprout to those that have nearly reached their full growth.

OPHIURANS, OR SERPENT-STARS.

The Serpent-Stars, or Ophiurans, are so called from the resemblance of their long slender rays to a snake's tail. They are found on nearly all coasts, and are at once distinguished by a small disk or central portion from which the rays start off very abruptly, instead of the gradual passage of the central part into the arms, as seen in the true Star-Fishes. They move about mainly by means of their spines. Nearly all have the arms simple, as seen in Figure 469, but some have the

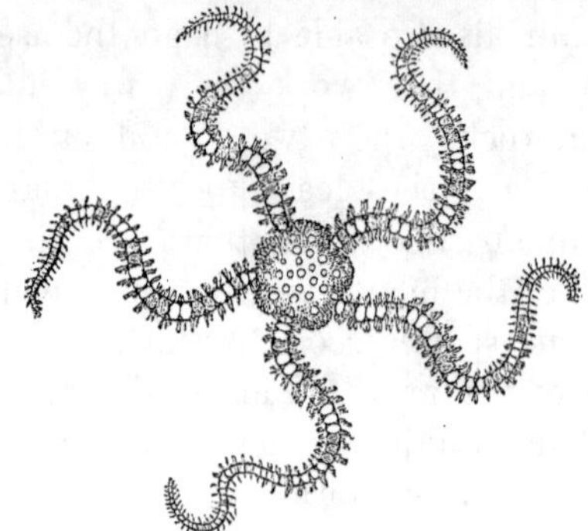

Fig. 469. — Serpent-Star, or Ophiuran.

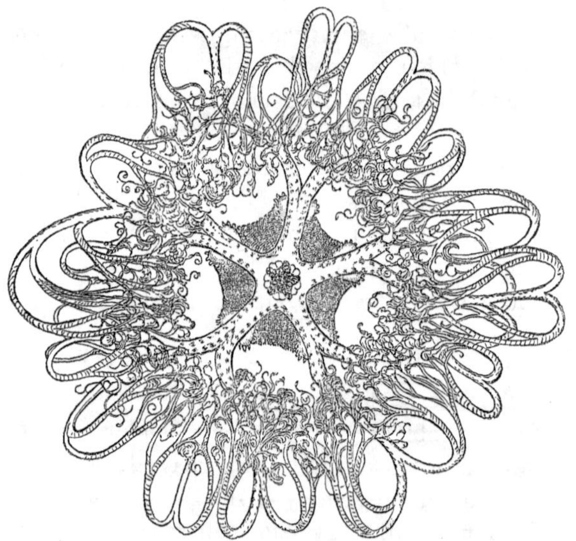

Fig. 470. — Basket-Fish, or Astrophyton.

arms much branched, as in the beautiful Astrophyton, Figure 470.

CRINOIDS.

The word Crinoid means *lily-like in form*, and is given to a large number of echinoderms on account of their lily-like or plant-like appearance. Only a small number of these animals is now living, and of the few living ones only one kind has a stem in the adult state, and this is the *Pentacrinus caput-medusæ*, of the West Indies, Figure 471. With the exception of this one, the living Crinoids much resemble

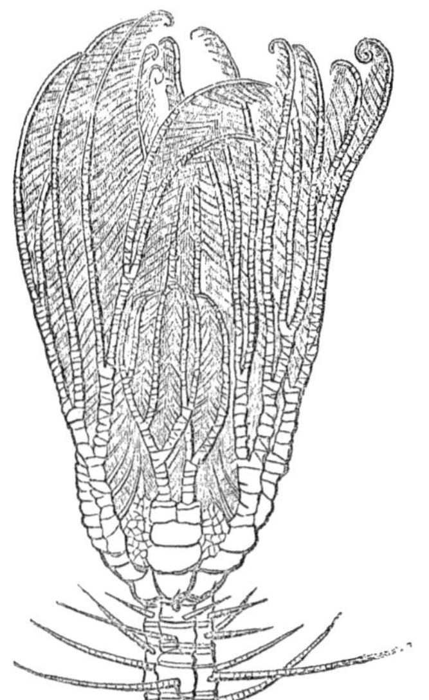

Fig. 471. — Living Crinoid.

the Star-Fishes and Ophiurans; but in the rocks, in various parts of the United States and in other countries, the stemmed kinds are exceedingly abundant, showing us that these animals lived in profusion in the old ocean which ages ago covered a large part of our country. And the fossil ones—as those found in the rocks are called—are so various in form, and so beautiful in patterns and markings, that no words can fitly describe them. The workman in the quarry stops to admire them, and the learned naturalist is fascinated by their beauty, and never grows weary of studying them. They are the "gems" of the geological collection, and their pictures are among the prettiest to be found in the Geological Reports. May every reader of these pages see at least one good collection of fossil Crinoids.

Of the living free Crinoids,—that is, those without a stem,—one of the best known is called the Comatula, or Feather-Star. When young this too has a stem, and looks not very unlike the Medusa's head, Figure 471; but as it grows older it drops from the stem, and lives a free life.

JELLY-FISHES, OR ACALEPHS.

Of all animals of the sea, perhaps none are more wonderful than these. Their jelly-like bodies, curious forms and structure, their beautiful colors of claret, rose, and pink, their varied and almost magical movements, as varied and graceful as those of the birds and insects of the air, their phosphorescence by night, causing them to be called the "Lamps of the Sea," and their curious changes in passing from the young to the adult state, have interested all intelligent visitors to the seaside, and have caused these animals to be carefully studied by some of the most eminent naturalists of Europe and America. The word Acaleph means *nettle*, and is given to these animals because some of them cause a stinging sensation when they touch our flesh; hence they are often called Sea-Nettles. They are also as often called Medusæ. Their

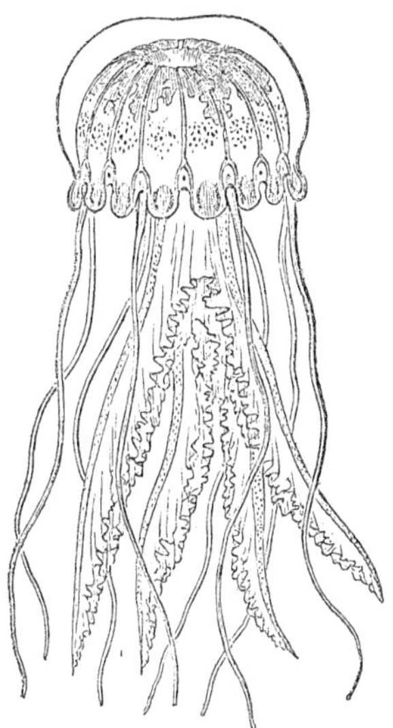

Fig. 472. — Jelly-Fish.

common name, Jelly-Fishes, was given on account of their jelly-like appearance and substance.

If we examine the structure of Acalephs, we find a cavity, which is the stomach, hollowed out of the mass of the body, and this cavity has an opening which serves as a mouth; the edges of this opening are turned outwards and prolonged into delicate fringes. And there are tubes which radiate from the centre of the body and unite with a tube at the circumference.

The kinds of Jelly-Fishes are numerous, and they vary in size from those scarcely visible to those which are one or two yards in diameter, and with tentacles thirty or forty feet long; and Mrs. Agassiz, in her beautiful book, "Seaside Studies," mentions one which measured about seven feet in diameter, and had tentacles more than a hundred feet in length!

Jelly-Fishes are a hungry race, and feed upon their own kind, and other marine animals, which they secure by means of their tentacles and *lassos*. On the tentacles of Jelly-Fishes, and of Polyps too, there are numerous lasso-cells, — too small to be seen without the microscope, — each containing a long, spirally-coiled thread or lasso, which can be instantly darted forth and fastened upon the little shrimp or other animal which is desired for food.

BEROID MEDUSÆ, OR CTENOPHORÆ.

The Beroid Medusæ are more or less spherical, or egg-shaped, with eight rows of locomotive fringes dividing the surface of the body as the ribs divide the surface of a melon. Pleurobrachia is one of the most common kinds on the northeast coast of the United States, and in its movements and curious appendages is one of the most wonderful of all the Medusæ. It is transparent, and besides the eight rows of fringes mentioned above, it has two most extraordinary tentacles, one on either side of the body; and no form of expansion or contraction, or curve or spiral, can be conceived of which these tentacles may not assume.

Bolina and Idyia are other ctenophoræ common on the northeast coast of the United States. The Rose-colored Idyia is three or four inches long, and shaped somewhat like a melon with one end cut off. The mouth occupies the whole of the cut-off end, and the digestive cavity, or stomach, occupies a large part of the interior of the animal. In summer it sometimes appears in such swarms as to tinge large patches

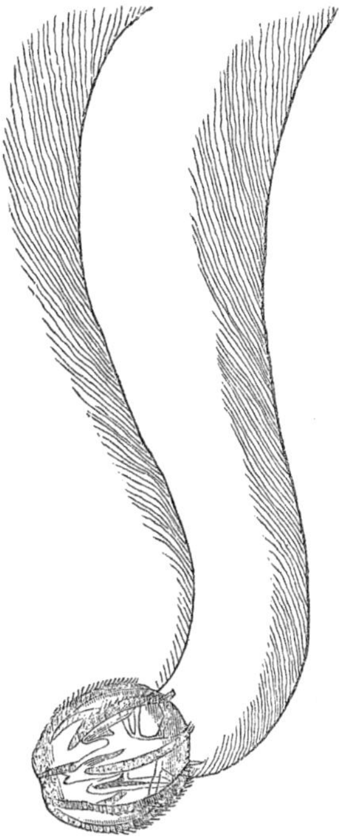

Fig. 473.—Pleurobrachia.

of the sea with a delicate rosy hue. It is very voracious, and feeds mainly on other jelly-fishes, sometimes capturing those nearly as large as itself.

TRUE MEDUSÆ, OR DISCOPHORÆ.

These have the body in the form of a hemispheric disk, more or less flattened. Of these disk-shaped medusæ none are more beautiful in their appearance or interesting in their history than the Aurelia, or "Sun-Fish," represented in Figure 477. This Jelly-Fish is

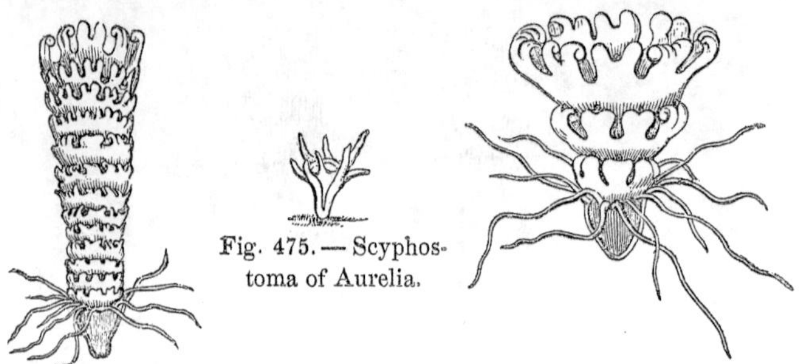

Fig. 474. — Strobila of Aurelia. Magnified.

Fig. 475. — Scyphostoma of Aurelia.

Fig. 476. — Strobila of Aurelia. Much magnified.

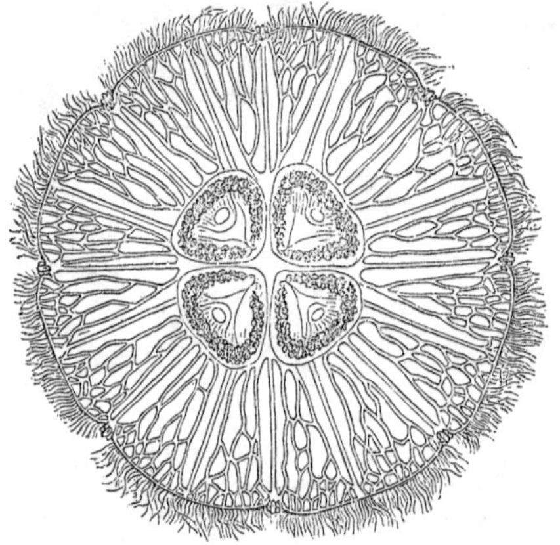

Fig. 477. — Sun-Fish, or Aurelia.

common on the coast of New England, and is about a foot across, in the larger specimens, and it lives but a single year. In the spring it is about a quarter of an inch in diameter, and on pleasant days moves in large swarms near the surface of the water. About the middle of summer they become full grown. Towards the close of summer they lay their eggs, and in the autumn they perish. At length the eggs hatch, and the little *planulæ*, as the newly hatched jelly-fishes are called, swim about in the water by means of little appendages which naturalists call *vibratile cilia.* Soon each becomes attached to a rock, shell, or sea-weed, and is then called *Scyphostoma*, Figure 475. Then the body begins to divide by horizontal constrictions, and soon appears as in Figures 474 and 476, and is then called *Strobila.* At length the segments become more and more separated, and the uppermost one drops off, then the next one, then the next, and so on till each in turn has separated from the one below itself. Each disk, as it separates, turns over and floats away, and is known as *Ephyra.* Soon each Ephyra assumes the form of a perfect jelly-fish, as shown in Figure 477. Thus one scyphostoma which comes from a single egg becomes a strobila, and this strobila divides into numerous parts, each of which becomes a jelly-fish.

HYDROIDS.

The Hydroids are jelly-fishes which are almost more wonderful in their mode of development than those already described. Occurring, as they do in many cases, in their early stages of existence, as mere discolored patches on sea-weeds, stones, or shells, or in appearance like little tufts of moss, or miniature shrubs, the untrained eye might well mistake the fact that they are animals. But naturalists have shown that these plant-like forms produce medusæ-buds, which expand into genuine medusæ or jelly-fishes. Figure 478 shows a little cluster of Hydroids attached to sea-weed, and Figure 479 shows a single individual of the

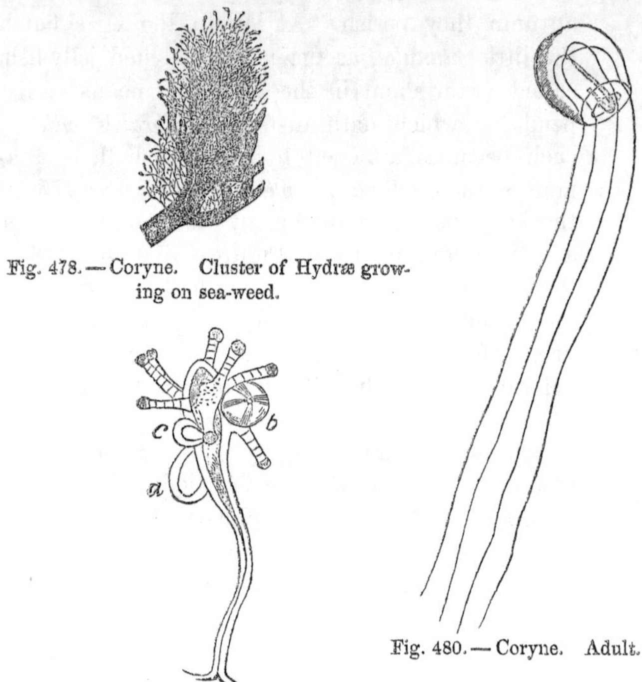

Fig. 478. — Coryne. Cluster of Hydræ growing on sea-weed.

Fig. 480. — Coryne. Adult.

Fig. 479. — Single individual of Fig. 478, enlarged, showing *a* and *b* just ready to drop off and become free medusæ, like Fig. 480; *c*, a younger bud.

same very much magnified, with two of the buds much enlarged, and a third quite prominent. Soon each bud becomes detached, and floats away as a free jelly-fish, like Figure 480, and is then known as Coryne, or, as it was formerly called, Sarsia, so named from Sars, a Norwegian naturalist, who was one of the first investigators of these curious kinds of jelly-fishes.

Nothing can excel the delicacy of Coryne. Soft as the softest jelly, almost as transparent as the dew-drop, yet it performs varied and rapid movements, contracts and expands its tentacles, catches and devours other medusæ, and other marine animals, and to all appear-

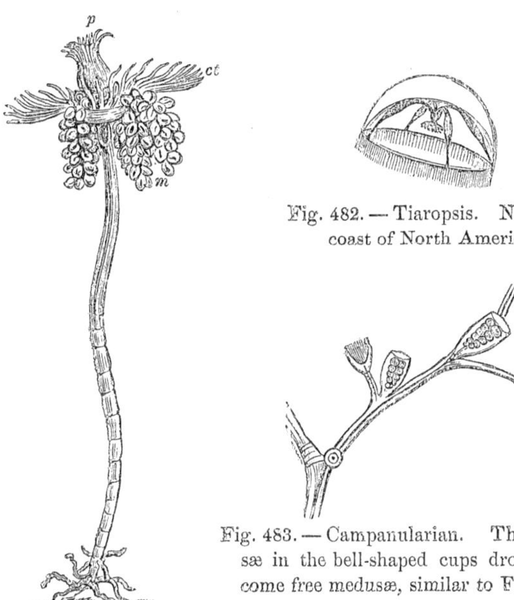

Fig. 482. — Tiaropsis. Northeast coast of North America.

Fig. 483. — Campanularian. The hydro-medusæ in the bell-shaped cups drop out and become free medusæ, similar to Fig. 482.

Fig. 481. — Tubularia. Massachusetts Bay.

m, medusæ; *ct*, coronal tentacle; *p*, proboscis.

ances delights in life as much as higher animals do. They are abundant in the spring. In the middle of summer they lay their eggs and perish. But the eggs do not hatch medusæ like the parent, but each hatches a little hydroid which is first free, then afterwards becomes attached to a shell, sea-weed, or stone, and from this

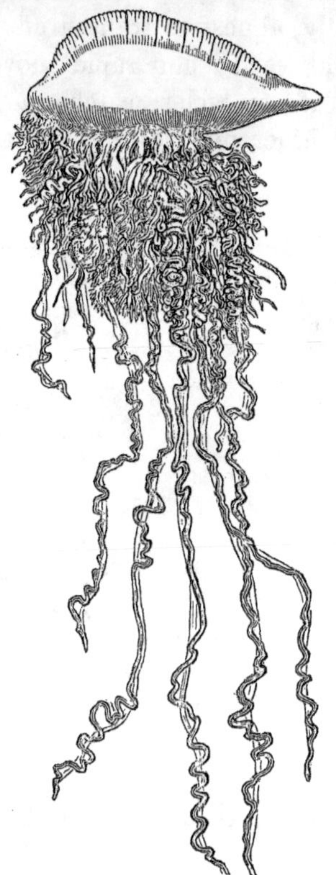

Fig. 484.—Portuguese Man-of-War.

little hydroid others branch till a little community of hydroids has grown up, as in Figure 478. From these hydroids bud again the Coryne, Figure 479.

In some kinds, as Tubularia, Figure 481, the hydroid

has a wreath of coronal tentacles, as they are called, a projecting part called a proboscis, and the medusæ grow in clusters from just above the coronal tentacles.

In those called Sertularians and Campanularians, Figure 483, the hydra has a stem which is covered by a horny sheath, forming a cup around the head. In a fertile cup there are a dozen or more hydro-medusæ, which at length drop out and become free medusæ similar to Tiaropsis, Figure 482.

In those called Siphonophoræ, the hydroid acalephs exist as free moving communities, each community being made up of individuals of different kinds, yet all so combined as to give the appearance of one animal. The Portuguese Man-of-War, of the Gulf of Mexico, is one of the most remarkable and best known of this sort. It consists of a pear-shaped and elegantly crested air-sac, floating lightly upon the water, and giving off from its under surface numerous long and varied appendages. These are the different members of the community, and fill different offices; some of them eat for the whole, others produce medusa-buds, and others are the locomotive or swimming members, and have tentacles that stretch out behind the floating community to the length of twenty or thirty feet.

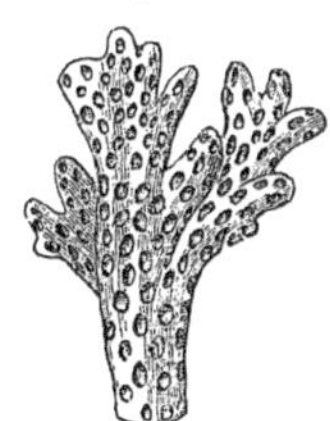
Fig. 484[a]. — Acalephian Coral.

It has recently been discovered by Professor Agassiz, that there are some kinds of Acalephs which produce coral similar to that formed by Polyps, described in the following pages, but unlike the latter in having, in the cells, a horizontal floor extending from wall to wall.

SEA-ANEMONES AND CORAL ANIMALS, OR POLYPS.

These are marine radiates which have a sack-like or

Fig. 485.—Polyp. A Sea-Anemone.

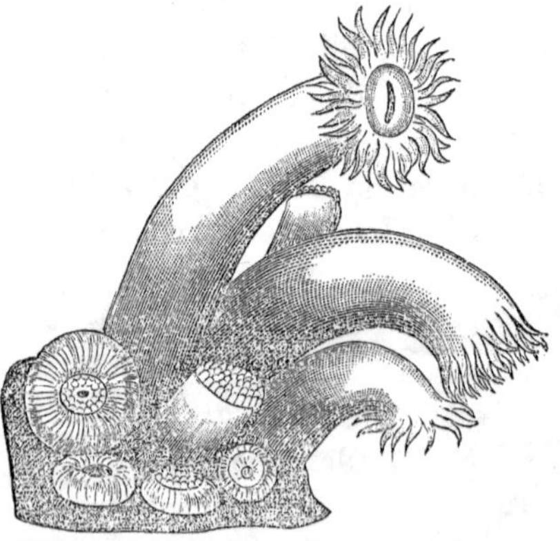

Fig. 486.—Cluster of Coral Polyps in various stages of expansion.

tubular body, with a circular top, in the centre of which

is an opening called the mouth, and around the mouth are one or more rows of hollow feelers, or tentacles. The mouth opens directly into an inner sack, which is the stomach, and this stomach opens at the bottom into the main body. The main body is divided by partitions, which run from the bottom to the top, and from the outer wall to the stomach. Through the opening at the bottom of the stomach there is free communication with all the chambers formed by the partitions, and these chambers connect with the tentacles; so that the food, after being digested, passes into the main body, and thence into the tentacles, thus nourishing every part. The food of polyps consists of small marine animals of various kinds, which are secured by means of the tentacles and the curious and wonderful lassos situated on the tentacles, and the nature of which has already been described on page 229. The word Polyp means *many-footed*, and is given to these animals on account of their numerous tentacles; but it must not be supposed that the latter are feet in any true sense. Most kinds of polyps are attached to the rocks, shells, or other bodies beneath the waves. Some live singly, others in communities whose numbers are often far more numerous than the leaves upon the trees. Polyps increase by means of eggs, by budding in a manner much like that of trees and shrubs, and by division of one animal into two or more, so that the largest communities arise from a single animal. The eggs are formed on the vertical partitions, and pass out, through the mouth, into the water. When first hatched the young do not look like the parent, but are little oval bodies which move freely about by means of the fringe-like appendages, called *vibratile cilia*, with which they are covered. At length each becomes attached to a rock, or shell, or sea-weed, and soon assumes the form of the parent. If it be a kind which buds, there soon grow from its sides or base others exactly like itself, and from these, in turn, bud other polyps of the same kind, and thus the community goes on growing till it has reached its limits of increase. If it be a kind which increases by division, it widens as it grows upward, and at length the walls in two opposite places begin to approach each other, and soon the polyp is divided into two, so that there are two mouths, and two circular disks surrounded by tentacles, instead of one as before the division; and the polyps thus formed divide in the same way, and this process is continued till from a single polyp there is formed a large and beautiful cluster.

Polyps readily reproduce lost parts, and even if cut in pieces, each fragment will, in some cases, become a perfect animal. Polyps vary in size from extreme minuteness to those that are more than a foot across. Some, like the Sea-Anemones, Figure 485, are wholly soft; others secrete a more or less solid framework, which is called *Coral;* and those which secrete coral are called Coral-Polyps, or Coral Animals. Some persons suppose that coral is something that is built by an insect, as the bee builds comb, or the wasp its nest, and the industry of this supposed insect is often spoken of. But it is not proper to give the name insect to the Coral-Polyps, for they are in no way related to insects, either in appearance, structure, or habits. Coral is not something which is built, but something which grows. It is the skeleton, or many united skeletons, of polyps, and these animals exhibit no industry in forming it, any more than do other animals in forming their own bones. Coral is not a house in which the animal lives; on the contrary, the coral is wholly inside of the animals, and it is only when the polyps die, wither, and disappear that we see the solid coral itself. Polyps grow in various and most wonderful and beautiful forms, imitating almost all kinds of vegetation, as lichens, fungi, mosses, ferns, grasses, herbs, shrubs, and trees. A hundred years ago, or more, they were thought to be plants, and even the great naturalist, Linnæus, regarded them as plant-animals, that is, partaking of the character of both plants and animals; but naturalists now regard them as true animals, although they are often called Zoöphytes, a word which means Animal-Plants. The colors of these wonderful animals of the sea are as beautiful and almost as varied as their forms; and some of the polyp communities equal, in splendor of colors, the most beautiful flower-gardens of the land; even beds of daisies, pinks, and asters have their rivals beneath the waves of the sea.

SEA-PENS, GORGONIAS, &c., OR ALCYONARIANS.

These are polyps which have eight long fringed or lobed tentacles, around a narrow disk, — Figures 487–489, — and which form compound clusters or communities by budding. The Sea-Pens, Verritillums, and Renillas are polyps which are arranged on a more or less expanded disk, which is connected with a sort of stem or peduncle, by means of which the community may move about or fix itself in the sand or mud. The Sea-Pens are so called from their resemblance to a quill. The Renilla, Figure 487, found on the coast

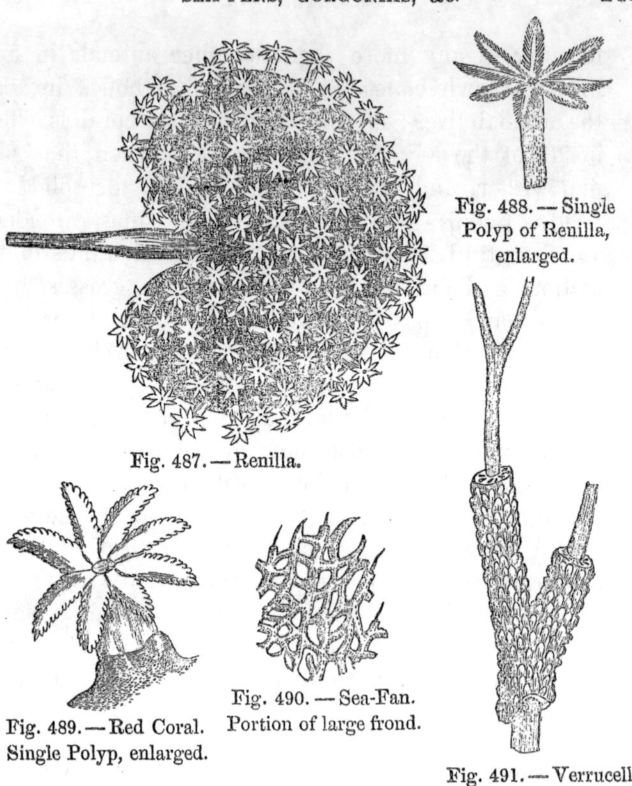

Fig. 488. — Single Polyp of Renilla, enlarged.

Fig. 487. — Renilla.

Fig. 489. — Red Coral. Single Polyp, enlarged.

Fig. 490. — Sea-Fan. Portion of large frond.

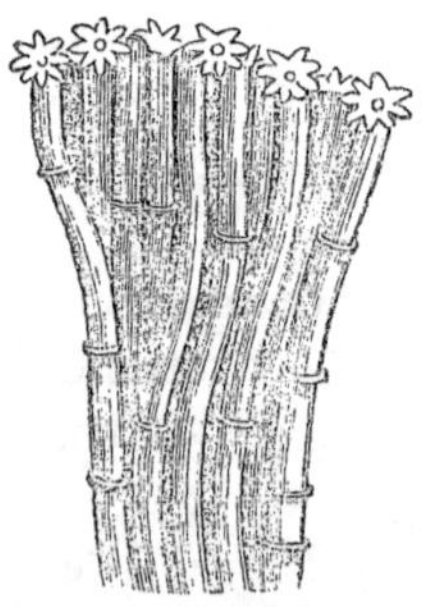

Fig. 491. — Verrucella.

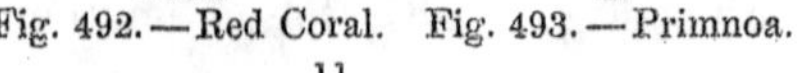

Fig. 492. — Red Coral.

Fig. 493. — Primnoa.

Fig. 494. — Organ-pipe.

of the Southern States and of South America, looks like a broad leaf attached to its leaf-stalk; and when the purple disk is covered with the expanded polyps, as seen in the cut, it is a very beautiful object. The exact form of the separate polyps is shown in Figure 488.

The Gorgonias abound in tropical seas, but some kinds are also found in temperate regions. The forms are exceedingly various, Figures 489–494, and many of them are very delicate and beautiful, often bearing a very close resemblance to plants; in all, however, the polyps are short, and secrete a solid central axis of coral. This axis is plainly shown in Figures 491 and 493. One of the most common and striking forms of the Gorgonias is the Sea-Fans, which are more or less broad and fan-shaped, the branches in many cases running together so as to form a network, Figure 490. One form of the Gorgonias, the Primnoa, Figure 493, is found even as far north as St. George's Banks and the Bay of Fundy. But the one which has the greatest popular interest is the Red Coral, *Corallium rubrum*. It is obtained mainly in the Mediterranean. The coral fishers go out in boats, and are provided with a large wooden cross, which is loaded with stone in the centre, and has hempen nets attached to each of its arms. While one man is constantly raising and letting fall this machine upon the coral beds, others row the boat so that the branches broken off are caught up by the nets. From time to time the cross and nets are raised, and the branches of coral which have been entangled in the meshes are secured.

Closely related to the Gorgonias are the Alcyonacea, of which the Organ-pipe Coral, Figure 494, is one of the most interesting examples. It is of a beautiful red color, and gets its name from the fact that the tubes of the coral somewhat resemble the pipes of an organ.

SEA-ANEMONES, OR ACTINARIANS.

These polyps are wholly soft, only a few secreting from the base a horn-like substance. They are common on nearly all coasts, and vary from a quarter of

Fig. 495. — Same as Fig. 498. Closed.

Fig. 496. — Same as Fig. 498. Just opening.

g. 497. — Sea-Anemone. Bunodes.

Fig. 498. — Sea-Anemone, or Fringed Actinia.

an inch to a foot or more in diameter, as seen in some of the tropical species. Our species seldom exceed two or three inches in diameter, and most of them are much smaller, although some are six inches high. The Bunodes, Figure 497, is found among the rocks on the coast of Maine. The most common kind on the northeast coast of North America is the Fringed Actinia, or Metridium, Figures 495, 496, 498. When fully expanded, it is about four inches high and three inches across the disk, and is a most interesting object for study.

MADREPORES, PORITES, MÆANDRINAS, ASTRÆAS, &c., OR MADREPORARIANS.

These polyps are simple or compound, often excessively branching, and they form coral in their walls, or outer parts, in their radiating partitions, and often at their base. The forms which the communities assume are very beautiful and exceedingly various, and are among the most beautiful objects in zoölogical cabinets.

The great group of Madrepores contains polyps which have a definite number of tentacles, twelve or more; those called Porites, Figure 500, have the cells shallow, and not more than one twelfth of an inch in diameter, and the coral in some cases branching, in others massive, and always very solid. Massive specimens of Porites are sometimes fifteen feet in diameter. In the true Madrepores, Figure 499, the polyps do not secrete coral at the base, and hence the cells of the coral are very deep, and these corals spread and branch into the most beautiful and varied forms, and the polyp at the end of a branch, Figure 499, is always larger than the others.

In the great group of the Astræans the tentacles occur in multiples of six. Those of this group, called Brain Corals, or Mæandrinas, have the surface covered

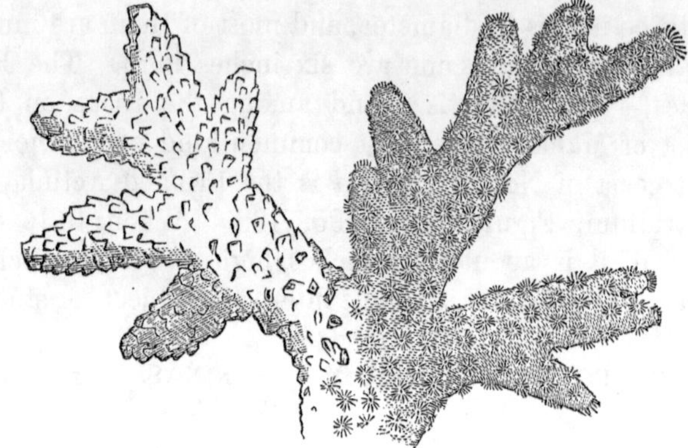

Fig. 499. — Madrepore. Right-hand branches alive.

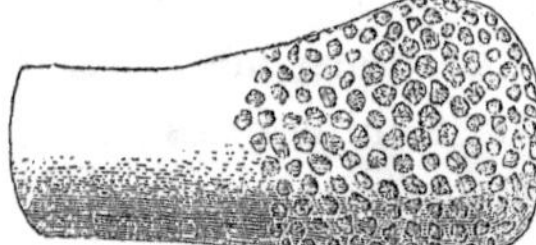

Fig. 500. — Porites.

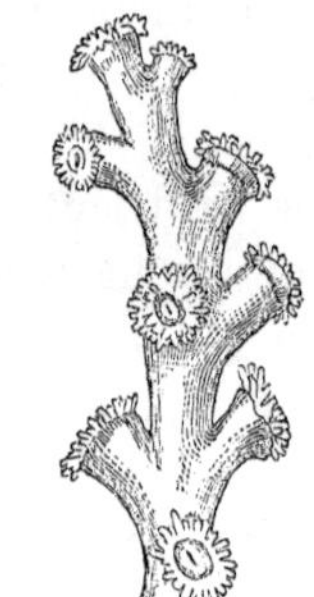

Fig. 501. — Cœnopsammia.

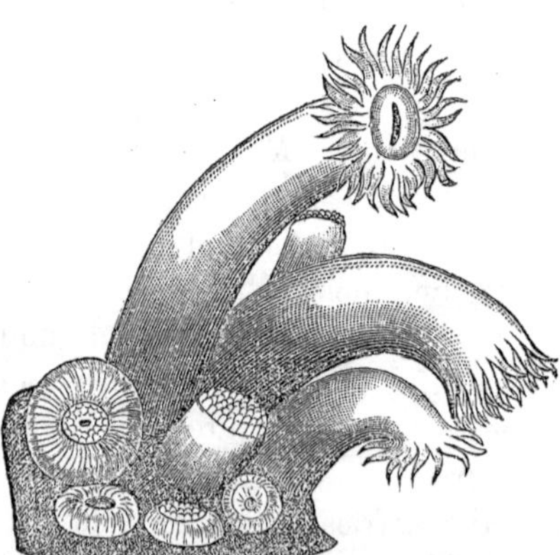

Fig. 502. — Astroides. Coral Polyps in various stages of expansion.

Fig. 503. — Dry Coral. Same as Fig. 502.

with winding trenches, Figure 504, on each side of which there is a row of tentacles. The form of the Mæandrinas is generally that of a hemisphere, and some of these masses are twelve feet in diameter. The true Astræans, or Star-Corals, Figure 506, have the cells in the form of concave pits, and the common forms of this coral are hemispherical or dome-shaped masses, some of which are twenty feet in diameter; and the polyps themselves are often an inch in diameter. Most of them, however, are very much smaller. One beautiful little Astræan, Dana's Astrangia, has its home in Long Island Sound, where it occurs in little clusters upon the stones and shells, from just below low-water mark even down to ten fathoms in depth. It thrives well in the aquarium, and eats little mollusks and other small animals with a good relish. In those coral polyps called Oculinas, the coral when young spreads so as to form a broad base; later beautiful tufts and tree-like branches arise from this base. A portion of one of these is shown in Figure 508.

In the great group of Fungus Corals, the coral is broad and flat, looking like a toad-stool without a stem, as in Figure 509. Polyps of this kind have short lobe-like tentacles in multiples of six. Each specimen, like Figure 509, is the secretion of a single polyp, and similar specimens are sometimes a foot or more in diameter.

But some of the most interesting facts about coral polyps remain to be told. Hundreds of the islands and reefs in the ocean are made of coral, — the skeletons of Polyps. These islands and reefs are most abundant and most extensive in the Pacific and Indian Oceans,

Fig. 504. — Mæandrina.

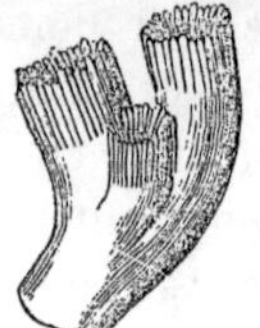

Fig. 505. — Cladocera.

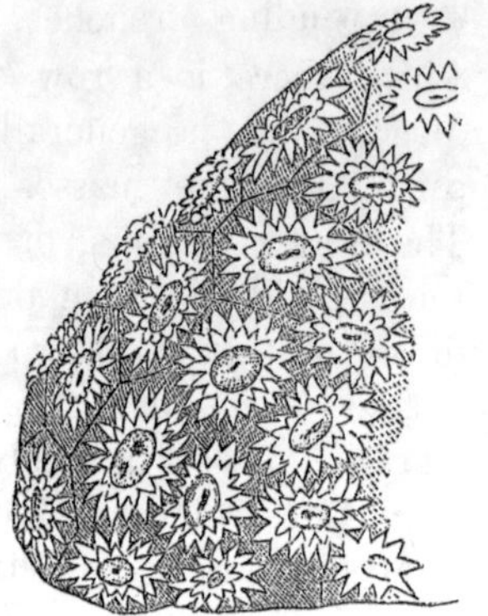

Fig. 506. — Star-Coral, or Astræa.

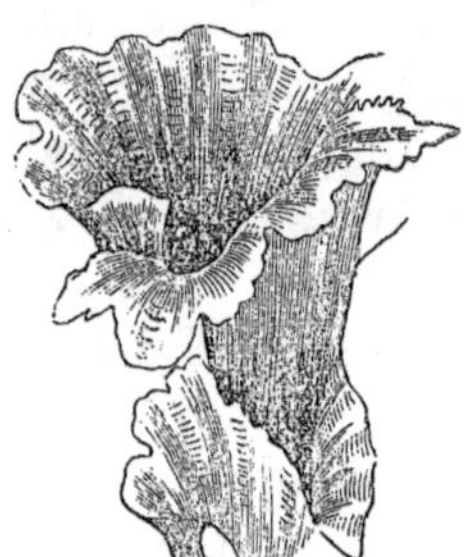

Fig. 507. —Merulina.

Fig. 508. — Oculina.

Fig. 509. — Fungus Coral.

but the islands which skirt the coast of Florida — the Keys — are also of coral formation, and according to Agassiz a large part of Florida itself is made of coral. Some reefs are small and have made only a little progress upward towards the surface of the water; others are miles in length and breadth, and come so near the surface of the water that it is dangerous for vessels to sail over them; and others still rise above the surface of the water forming islands which, in some cases, are covered with coral sand, and in others with a more or less luxuriant growth of tropical vegetation. Reefs stretch north and south near New Caledonia for the distance of four hundred miles, and along the north-eastern coast of Australia for a thousand miles. When a reef or bank of coral is near the shore, it is called a Fringing Reef; when at a distance from the shore, a Barrier Reef; and when it surrounds a body of water, as is often the case in the Pacific, an Atoll or Coral Island. The corals which form the principal part of the reefs and islands are Madrepores, Porites, Mæandrinas, and Astræas; the frailer corals, such as the Sea-Fans and other Gorgonias, adorn the reef as it nears the surface of the water, but do not contribute much to its growth.

From what has already been said, it is hoped that it will be understood that the reefs and islands are not something which the coral animals build, as a mason builds a house, or as a bee or wasp builds her nest or comb, but that the reefs and islands are made up of the hard parts or skeletons of polyps that lived and died where the reef or island now stands.

Only about an inch of a growing coral mass or reef is alive, all the rest within is dead; death goes on below as fast as growth goes on above. When the reef at last grows up to the surface of the water, the polyps die; for they cannot live out of water. The winds and waves do the rest; they break fragments from the sides of the reef and pile them nearer the centre; they bring sea-weeds and other floating materials, and cast them over the whole; plants at length spring up, and in the course of years the island — except its broad beaches of coral sand — is clothed with verdure, and man, perhaps, comes there and makes his home. These little polyps, then, are increasing the amount of dry land on the surface of the globe; and in this and in other ways God makes their lives serve great and important ends.

But a history of the polyps would be unfinished if we should not mention their connection with some of the rocks of the globe, — the limestones. It is a very interesting fact that reef corals and limestone, or marble, have essentially the same chemical composition; and it is well known that some of the coral reefs of the Pacific, which have been lifted out of water by volcanic forces, are nearly or quite as solid as ordinary marble. From these facts, and many others, geologists believe that a large part of the limestones of the globe are made out of the coral reefs that grew in the old oceans, which long before the creation of man covered the countries where marble is now found. If this be true, many of the rocks which underlie vast countries, the marble temples and palaces of the East, the marble monuments and public buildings of our own country, the mortar upon the walls and ceilings of our houses, and the marble tables and mantel-pieces so highly prized, have all come from the skeletons of these little flower-like animals of the sea. Their skeletons have furnished even the blocks of marble which the sculptor chisels, and are thus inseparably linked with the highest department of culture and of art in which the mind and hand of man can engage.

PROTOZOANS.

THERE is a vast number of beings which are so simple in their structure that naturalists are in doubt, in many cases, whether to call them Plants or Animals. These are now called Protozoans, a word which means *first* or *simplest animals*. A few of the forms are shown in Figures 510–520, — all much enlarged, except Figures 518, 519, 520. In most cases they have neither mouth nor stomach, and, excepting the Sponges and some others, are exceedingly minute and mostly microscopic. They are doubtless more numerous than all the other animals of the globe, for they live in immense numbers in every ditch and pool, every stream, pond, and lake, and in almost every part of the sea. There is scarcely a drop of water that is not inhabited by some of them. They were also exceedingly abundant in the past ages of the world; for their skeletons, or hard parts, fill the rocks in many places, and rocky strata hundreds of feet in thickness are wholly made up of their remains.

One group of the Protozoans is called Infusoria, from having first been found in vegetable infusions, that is, in liquids in which plants have been immersed; of these, Vorticella, Figure 510, is a well-known kind.

There is another group called Rhizopods — a word

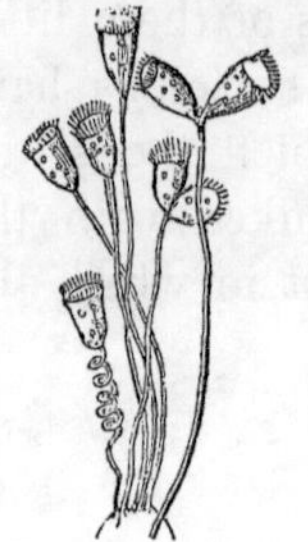

Fig. 510.—Vorticella.

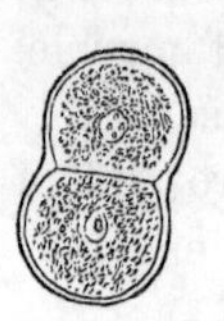

Fig. 511.—Gregarina.

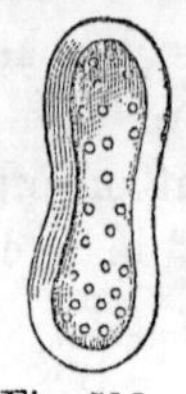

Fig. 512.—Sphærozoum.

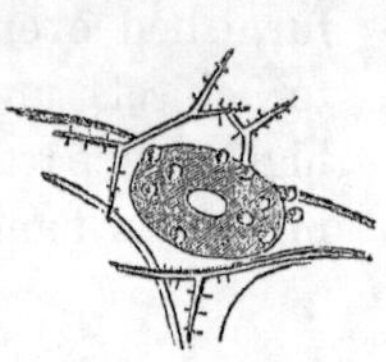

Fig. 513.—Portion of Fig. 512, magnified.

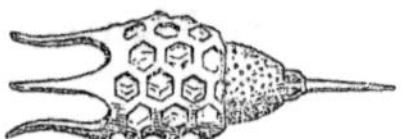

Fig. 514.—Podocyrtis.

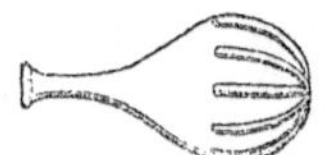

Fig. 515.—Lagena.

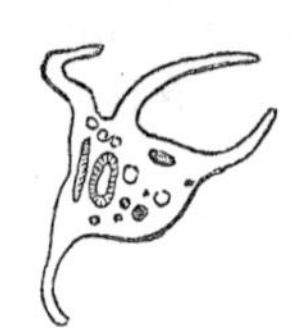

Fig. 516.—Amœba.

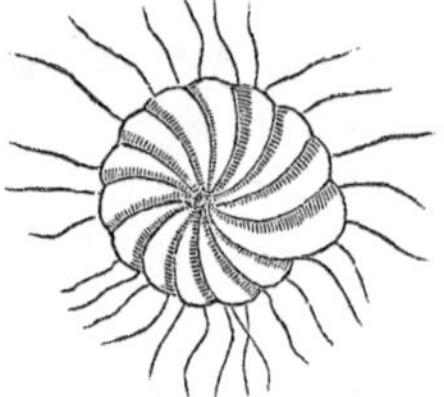

Fig. 517.—Polystomella.

Fig. 518.—Nummulite.

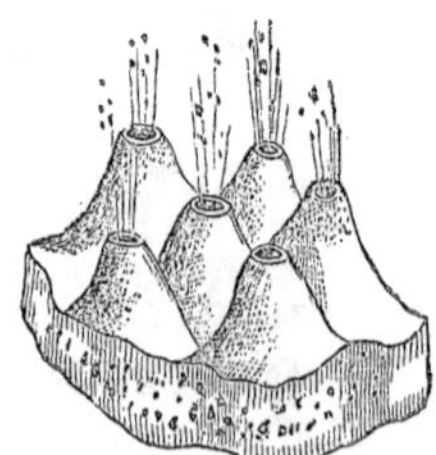

Fig. 519.—Sponge, alive.

Fig. 520.—Sponge.

Protozoans.

meaning *root-feet* — because they throw out fibre or root-like appendages, as in Figures 516, 517. Most of these have a shell, and are often called Foraminifers, from the pores or foramens in the shell, through which the appendages just mentioned are thrust out. The vast chalk-beds of Europe are almost wholly made of the shells of Rhizopods, which are so minute that a million are contained in a cubic inch of the chalk. The Nummulite, Figure 518, is one of the Rhizopods or Foraminifers, which has a shell half an inch or more in diameter, in some cases, and divided into chambers which resemble those of a Nautilus or Ammonite. Extensive beds of limestone are made of Nummulites; that of which the Pyramids of Egypt are built is filled with shells of this sort. The Amœba, Figure 516, is a Rhizopod which has no shell. It is a simple, almost fluid mass, seen only by the aid of a microscope, and it changes its form almost every moment. It has neither mouth nor stomach, yet on coming to a particle of food it readily closes around it, and digests it, any part of the body being formed into mouth, stomach, or tentacles, as the occasion requires!

Sponges are protozoans which have been regarded by many as plants, but are now generally considered to be compound animals. They are common in ponds and lakes, as well as in nearly all parts of the sea, and their forms are exceedingly various and often extremely beautiful. Some cover the rocks like a carpet of mosses; others grow in massive clusters; others branch like trees and shrubs; and others still take the form of the most elegant cups, goblets, and vases.

CONCLUSION.

In these few pages we have endeavored to make you acquainted with some of the principal forms in which animals have been created, and thus give you some idea of the Animal Kingdom. Although only a few kinds out of the many thousands now living have been mentioned, you have learned that all the Animals upon our globe may be divided into four, or at most five, great groups, — the Vertebrates or Backboned Animals, the Articulates or Jointed Animals, the Mollusks or Soft-bodied Animals, the Radiates or Star-shaped Animals, and perhaps a fifth group called the Protozoans; and it may be added that geologists tell us that all the animals of past ages, which are now known only by their remains, but which were so numerous that in many places they fill the rocks to the depth of miles, also belong to either one or the other of these five groups. Naturalists call these groups Branches. You have learned that the Vertebrates are divided into Mammals, Birds, Reptiles, Batrachians, and Fishes; that the Articulates are divided into Insects, Crustaceans, and Worms; that the Mollusks are divided into Cephalopods, Gasteropods, and Acephals; and that the Radiates are divided into Echinoderms, Acalephs, and Polyps. Naturalists call these groups Classes. You have learned that the Mammals are divided into Man, Monkeys, Carnivores or Beasts of Prey, Herbivores or Plant-eaters, Cetaceans or Whales, Bats, Insect-eaters, Rodents or Gnawers, Edentates and Marsupials; that the Birds are divided into Birds of Prey, Climbers, Perchers, Scratchers, Runners, and Swimmers; and that the Reptiles, Batrachians, and Fishes, the Insects, Crustaceans, and Worms, the Cephalopods, Gasteropods, and Acephals, and the Echinoderms, Acalephs, and Polyps, are also similarly divided into groups. Naturalists call these groups Orders. The Orders are divided into Families, — for example, the Order of Birds of Prey is divided into the Family of Vultures, the Family of Falcons and Eagles, and the Family of Owls. Families are divided into Genera, — for example, the Family of Falcons is divided into true Falcons, Hawks, Eagles, &c. Genera are divided into Species, — for example, the Genus of true Falcons is divided into the Peregrine Falcons or Duck Hawks, Pigeon Falcons or Pigeon Hawks, Sparrow Falcons or Sparrow Hawks, &c.

You have gained some idea of the way in which animals are distributed over the surface of the globe. Each zone of the earth's surface, each zone of height, each hemisphere, each grand division of the earth, has its own kinds of animals; even each of the different parts of every country has animals peculiar to itself. And it is so in the waters; each ocean and sea, each gulf and bay, and each zone of depth, has its own animal forms, such as are found nowhere else.

But the words of a book cannot fitly describe the living beings of our globe. We need to open our eyes and study them in the world about us. We may find them everywhere, — in forest and field, on the mountain and in the sea, in every stream, pond, and lake, in every pool and ditch and bog, and in every glass of water from the spring. Every summer's day brings scores of beautiful winged forms, and on every summer's night other not less beautiful forms flit about our lamps, or look in at our windows, tempting us to study and admire them. And how full of interest is every one of them, whether it be the Deer bounding through the forest or over the plain, the Eagle soaring above our heads until lost amid the clouds, the Butterfly flitting from flower to flower, the Mussel plowing its way in the river's sand, or the little Polyp beneath the ocean wave. And they are interesting not merely on account of their varied and beautiful forms and colors, wonderful structure, and often marvellous instincts and habits, and great variety of uses, but because they are the works of God,—His thoughts expressed in visible forms. If we study these wonderful objects in the right spirit, we shall learn more of Him who made them, and who careth for them,—suffering not even a sparrow to fall without His notice.

INDEX.

Q

Y.

Z.

APPENDIX.

Those who desire to learn to mount and preserve Mammals, Birds, and other animals will do well to procure "The Taxidermist's Manual," by S. H. Sylvester, Middleboro, Mass.

Artificial eyes of all sizes and colors can be obtained of C. F. A. Hinrichs, Broadway, New York.

CLASSIFICATION OF ANIMALS.

THE BRANCH OF VERTEBRATA, OR VERTEBRATES.

THE CLASS OF MAMMALIA, OR MAMMALS.

ORDERS.	FAMILIES.	*Genera.*	
BIMANA or MAN.	HUMAN FAMILY.	*Homo*	or Man.
QUADRUMANA or MONKEYS.	SIMIADÆ or OLD WORLD MONKEYS.	*Troglodytes*	or Chimpanzee & Gorilla.
		Simia	" Orang-Outang.
		Hylobates	" Gibbons.
		Semnopithecus	" Solemn Apes.
		Presbytis	" Tailed Gibbons.
		Cercopithecus	" Guenons.
		Colobus	" Thumbless Apes.
		Macacus	" Macacos.
		Inuus	" Barbary Ape.
		Cynocephalus	" Baboons.
	CEBIDÆ or NEW WORLD MONKEYS.	*Mycetes*	or Howlers.
		Ateles	" Spider Monkeys.
		Lagothrix	" Glutton "
		Cebus	" Weepers.
		Pithecia	" Fox-tailed Monkeys.
		Callithrix	" Squirrel "
		Nyctipithecus	" Night "
		Jacchus and Midas	" Marmosets.
	LEMURIDÆ or LEMURS.	*Lemur*	or true Lemurs or Makis.
		Indris	" Indri.
		Loris	" Lorises.
		Galago	" Galagos.
		Tarsius	" Tarsiers.
		Chiromys	" Aye-Aye.
CARNIVORA or FLESH-EATERS.	FELIDÆ or CAT FAMILY.	*Felis*	or Lions, Tigers, Panthers, Leopards, Puma, Cat.
		Lynx	" Wild Cats, Lynx, &c.
	HYENADÆ	*Hyena*	or Hyenas.
	CANIDÆ.	*Canis*	or Wolves, Jackals, Dogs.
		Vulpes	" Foxes.
	VIVERRIDÆ or CIVET FAMILY.	*Bassaris*	or Civet Cat.
		Viverra	" Civets and Genet.
		Herpestes	" Ichneumons.
	MUSTELIDÆ or WEASEL FAMILY.	*Mustela*	or Martens or Sable.
		Putorius	" Weasels and Minks.
		Gulo	" Wolverines.
		Lutra	" Otters.
		Mephitis	" Skunks.
		Taxidea	" Badgers.
	URSIDÆ or BEAR FAMILY.	*Procyon*	or Raccoons.
		Ailurus	" Panda.
		Ursus	" Bears.
	PHOCIDÆ or SEAL FAMILY.*	*Phoca, &c.*	or Seals.
		Rosmarus	" Walrus or Morse.

* According to Professor Gill, the old family Phocidæ really comprises three families: PHOCIDÆ proper, including *Phoca*, *Pagomys*, *Pagophilus*, *Erignathus*, *Halichœrus*, *Monachus*, *Cystophora*, *Macrorhinus*, *Lobodon*, *Stenorhynchus*, *Leptonyx* and *Ommatophora*; OTARIIDÆ, including *Otaria*, *Callorhinus*, *Eumetopias*, *Zalophus*, and *Arctocephalus*; and ROSMARIDÆ, including *Rosmarus*.

VERTEBRATES: MAMMALS — *Continued.*

ORDERS.	FAMILIES.	Genera.	
HERBIVORA or PLANT-EATERS.	ELEPHANTIDÆ or ELEPHANT FAMILY.	*Elephas* *Mastodon*	or Elephants. (Fossil.)
	RHINOCEROTIDÆ.	*Rhinoceros*	or Rhinoceroses.
	TAPIRIDÆ or TAPIR FAMILY.	*Tapirus*	or Tapirs.
	HYRACIDÆ.	*Hyrax*	or Damans.
	SUIDÆ or HOG FAMILY.	*Sus* *Phacochœrus* *Dicotyles*	or Wild Boar, &c. " Wart-bearing Hogs. " Peccaries.
	HIPPOPOTAMIDÆ	*Hippopotamus*	or " River Horse."
	EQUIDÆ.	*Equus*	or Horse, Ass, Zebra, &c.
	CERVIDÆ or DEER FAMILY.	*Alce* *Rangifer* *Cervus*	or Moose & European Elk. " Reindeer & Caribou. " Common Deer, Wapiti, Stag, &c.
	ANTILOPIDÆ or HOLLOW-HORNED RUMINANT FAMILY.	*Antilope* *Antilocapra* *Aplocerus* *Capra* *Ovis* *Ovibos* *Bos*	or Antelopes. " Pronghorn. " Rocky Mountain Goat. " Goats. " Sheep. " Musk Ox. " Oxen, Buffaloes, &c.
	CAMELOPARDALIDÆ.	*Camelopardalis*	or Giraffe.
	CAMELIDÆ.	*Camelus* *Auchenia*	or Camels. " Llamas.
	MOSCHIDÆ.	*Moschus*	or Musk Deer.
	SIRENIDÆ.	*Manatus* *Halicore* *Rytina*	or Sea-Cows. " Dugong. " Stellers.

[VERTEBRATES: MAMMALS — *Continued*.]

ORDERS.	FAMILIES.	*Genera.*		
MUTILATA or CETACEA or WHALES, &c.	BALÆNIDÆ or RIGHT WHALE FAMILY.	*Balæna*	or	Greenland Whale, or Right Whale.
		Balænoptera	"	Rorquals or Finners.
	PHYSETERIDÆ or CATODONTIDÆ.	*Physter*, or *Catodon*	or	Sperm Whales.
	DELPHINIDÆ or DOLPHIN FAMILY.	*Beluga*	or	White Whale or White Grampus.
		Globicephalus	"	Bottleheads.
		Phocæna	"	Porpoises & Grampuses.
		Delphinus	"	Dolphins.
		Delphinorhynchus	"	Beaked Dolphins.
		Delphinapterus	"	Peron's Dolphin.
		Soosoo	"	Soosoos of the Ganges.
		Inia		
	MONODONTIDÆ.	*Monodon*	or	Narwhal.
CHEIROPTERA or BATS.	PTEROPODIDÆ or FRUGIVOROUS BAT FAMILY.	*Pteropus*	or	Rousettes.
	MEGADERMATIDÆ or HORSE-SHOE BAT FAMILY.	*Rhinolophus*	or	Horse-shoe Bats.
		Megaderma	"	Megaderms.
		Macrotus		
		&c.		
	PHYLLOSTOMATIDÆ.	*Phyllostoma*	or	Vampire Bats.
	NOCTILIONIDÆ.	*Nyctinomus, &c.*	or	Noctilios, &c.
	VESPERTILIONIDÆ or COMMON BAT FAMILY.	*Nycticejus*		
		Lasiurus	or	Red and Hoary Bats.
		Scotophilus	"	Carolina Bat, &c.
		Vespertilio	"	Little Brown Bat, &c.
		Synotus	"	Big-eared Bats.
		Antrozous	"	Pale Bat.
INSECTIVORA or INSECT-EATERS.	DERMOPTERA.	*Galeopithecus*	or	Galeopithecus.
	SCANDENTIA.	*Cladobates, &c.*	or	Banxrings.
	SORICIDÆ or SHREW FAMILY.	*Neosorex & Sorex*	or	Shrews.
		Blarina	"	Mole Shrews.
	TALPIDÆ or MOLE FAMILY.	*Scalops*	or	Shrew Moles.
		Condylura	"	Star-nosed Moles.
		Talpa	"	European Mole.
		Urotrichus	"	Bulb-nosed Mole.
		Chrysochloris	"	Golden-green Moles.
	ACULEATA or HEDGEHOG FAMILY.	*Erinaceus*	or	Hedgehogs.
		Centetes	"	Tenrecs.

[VERTEBRATES: MAMMALS — *Continued.*]

ORDERS.	FAMILIES.	*Genera.*	
RODENTIA or GNAWERS.	SCIURIDÆ or SQUIRREL FAMILY.	*Sciurus*	or True Squirrels.
		Pteromys	" Flying Squirrels.
		Tamias	" Striped Squirrels.
		Spermophilus	" Spermophiles or Gophers.
		Cynomys	" Prairie Dogs.
		Arctomys	" Woodchucks & Marmots.
		Myoxus	" Dormice.
		Castor	" Beavers.
		Aplodontia	" Sewellel.
	SACCOMYIDÆ or POUCHED GOPHER FAMILY.	*Geomys* and *Thomomys*	or Pouched Gophers.
		Dipodomys	" Kangaroo Rats.
		Perognathus	" Tuft-tailed Mice.
	MURIDÆ or RAT FAMILY.	*Dipus*	or Jerboas.
		Jaculus	" Amer. Jumping Mouse.
		Gerbillus	" Gerbils.
		Mus	" Rats.
		Cricetus	" Hamsters.
		Reithrodon	" Harvest Mice.
		Hesperomys	" White-footed Mice.
		Neotoma	" Wood Rats.
		Sigmodon	" Cotton Rats.
		Arvicola	" Field Mice.
		Myodes	" Lemmings.
		Fiber	" Muskrat.
	HYSTRICIDÆ or PORCUPINE FAMILY.	*Erethizon* & *Hystrix*	or Porcupines.
		Dasyprocta	" Agoutis.
		Dolichotis	" Patagonian Cavies.
		Chinchilla	" Chinchillas.
		Cavia	" Guinea Pigs.
		Myopotamos	" Couia.
		Hydrochœrus	" Capybara.
	LEPORIDÆ.	*Lepus*	or Hares and Rabbits.
		Lagomys	" Pikas.
EDENTATA or EDENTATES.	BRADYPODA.	*Bradypus*	or Sloths.
		Megatherium. Megalonyx, & Mylodon.	Extinct.
	EFFODIENTA.	*Dasypus, &c.*	or Armadillos, &c.
		Glyptodon	(Extinct.)
		Myrmecophaga, &c.	" Ant-eaters, &c.
MARSUPIALIA or MARSUPIALS.	PHALANGISTIDÆ.	*Phalangista, &c.*	or Phalangers, &c.
	DASYURIDÆ.	*Dasyurus, &c.*	or Bear Opossums, &c.
	MACROPODIDÆ.	*Macropus, &c*	or Kangaroos, &c.
	PERAMELIDÆ.	*Perameles*	or Bandicots, &c.
	DIDELPHIDÆ.	*Didelphys, &c.*	or Opossums, &c.
	PHASCOLOMYIDÆ.	*Phascolomys*	or Wombat.
MONOTREMATA.	PLATYPUS or DUCKBILL FAMILY.	*Ornithorhynchus* or *Platypus*	or Duckbill.
		Echidna	" Porcupine Ant-eater.

THE CLASS OF BIRDS.

ORDERS.	FAMILIES.	*Genera.*	
RAPTORES or BIRDS OF PREY.	VULTURIDÆ or VULTURE FAMILY.	*Vultur*	or Condor, &c
		Gypætos	" Læmmergyer.
		Cathartes	" North Amer. Vultures.
	FALCONIDÆ or FALCON FAMILY.	*Falco*	or Falcons.
		Astur	" Goshawk, &c.
		Accipiter	" Cooper's Hawk, &c.
		Buteo	" Buzzards.
		Archibuteo	" Rough-legged Hawk.
		Astarina	
		Nauclerus	" Swallow-tailed Hawk.
		Elanus	" White-tailed Hawk.
		Ictinia	" Mississippi Kite.
		Rostrhamus	" Black Kite.
		Circus	" Marsh Hawk.
		Aquila	" Golden Eagle.
		Halietus	" White-headed and Sea Eagles.
		Pandion	" Fish Hawks.
		Polyborus	" Caracara Eagle.
		Craxirex	
	STRIGIDÆ or OWL FAMILY.	*Strix*	or Barn Owls.
		Bubo	" Great Horned Owls.
		Scops	" Screech Owls.
		Otus	" Long-eared Owls.
		Brachyotus	" Short-eared Owls.
		Syrnium	" Gray Owls.
		Nyctale	" Sparrow Owls.
		Athene	" Burrowing Owls.
		Glaucidium	" Pigmy Owls.
		Surnia	" Day Owls.
SCANSORES or CLIMBERS.	PSITTACIDÆ.	*Conurus*, &c.	or Parrots.
	RHAMPHASTIDÆ.	*Rhamphastos*	or Toucans.
	TROGONIDÆ.	*Trogon*	or Trogons.
	CUCULIDÆ or CUCKOO FAMILY.	*Crotophaga*	or Black Parrot and Ani.
		Geococcyx	" Road Runner.
		Coccygus	" Cuckoos.
	PICIDÆ or WOODPECKER FAMILY.	*Campephilus*	or Ivory-billed Woodpeckers.
		Picus	" Hairy and downy Woodpeckers.
		Picoides	" Three-toed Woodpeckers.
		Sphyrapicus	" Yellow-bellied Woodpecker, &c.
		Hylatomus	" Black Woodcock.
		Centurus	" Red-billed Woodp., &c.
		Melanerpes	" Red-headed Woodpeckers.
		Colaptes	" Golden-winged Woodp.

[VERTEBRATES: BIRDS — *Continued.*]

ORDERS.		FAMILIES.	*Genera.*		
INSESSORES or PERCHERS.	STRISORES.	TROCHILIDÆ.	*Trochilus, &c.*	or	Humming-Birds.
		CYPSELIDÆ.	*Chætura, &c.*	or	Chimney Swallows and Swifts.
		CAPRIMULGIDÆ.	*Antrostomus*	or	Chuckwill's Widow and Whippoorwills.
			Chordeiles	"	Night-Hawks.
	CLAMATORES.	ALCEDINIDÆ.	*Ceryle*	or	King-fishers.
		PRIONITIDÆ.	*Momotus*	or	Saw-Bills.
		COLOPTERIDÆ or FLYCATCHER FAMILY.	*Pachyrhamphus*	or	Rose-throated Flycatchers.
			Milvulus	"	Forked-tailed Flycatchers.
			Tyrannus	"	King-birds, &c.
			Myiarchus	"	Great-crested Flycatchers, &c.
			Sayornis	"	Phœbe Bird or Pewee,&c
			Contopus	"	Wood Pewee, &c.
			Empidonax	"	Least Flycatcher, &c.
			Pyrocephalus	"	Red Flycatcher.
	OSCINES.	TURDIDÆ or THRUSH FAMILY.	*Turdus*	or	Wood & Hermit Thrushes, Robin, &c.
			Saxicola	"	Stone Chats.
			Erythaca	"	Robin Redbreast.
			Sialia	"	Blue-Birds.
			Regulus	"	Ruby-crowned Wren,&c.
			Hydrobata	"	Water Ouzels.
		SYLVICOLIDÆ or WARBLER FAMILY.	*Philomela*	or	Nightingales.
			Anthus	"	Tit Larks.
			Neocorys	"	Missouri Skylark.
			Mniotilta	"	Black & White Creepers.
			Parula	"	Blue Yellow-backed Warblers.
			Protonotaria	"	Prothonotary Warblers.
			Geothlypis	"	Maryland Yellow-throat, &c.
			Oporornis	"	Connecticut Warbler,&c.
			Icteria	"	Chats.
			Helmitherus	"	Worm-eating Warbler, &c.
			Helminthophaga	"	Golden-winged Warbler.
			Seiurus	"	Golden-crowned Thrush, &c.
			Dendroica	"	Yellow-rumped Warbler, Blackburnian, &c.
			Myiodioctes	"	Hooded Warbler, &c.
			Cardellina	"	Vermilion Flycatcher.
			Setophaga	"	Redstarts.
			Pyranga	"	Tanagers.
		HIRUNDINIDÆ or SWALLOW FAMILY.	*Hirundo*	or	Swallows.
			Cotyle	"	Bank Swallows.
			Progne	"	Purple Martins.
		BOMBYCILLIDÆ.	*Ampelis*	or	Wax-wings.
			Myiadestes	"	Townsend's Flycatcher.

[VERTEBRATES: BIRDS — *Continued.*]

ORDERS.		FAMILIES.	Genera.	
INSESSORES or PERCHERS — *Continued.*	OSCINES — *Continued.*	LANIDÆ.	*Collyrio*	" Shrikes.
			Vireo	" Vireos.
		LIOTRICHIDÆ or MOCKING-BIRD FAMILY.	*Mimus, &c.*	or Mocking and Cat Birds.
			Harporhynchus	" Brown Thrushes.
			Catherpes	" White-throated Wren.
			Salpinctes	" Rock Wren.
			Thryothorus	" Gt. Carolina & Bewick's.
			Cistothorus	" Long-billed Marsh Wren.
			Troglodytes	" House Wren, &c.
		CERTHIADÆ.	*Certhia*	or Creepers.
			Sitta	" Nuthatches.
		PARIDÆ.	*Polioptila*	or Blue Gray Flycatcher.
			Lophophanes	" Tufted Titmice.
			Parus, &c.	" Titmice.
		DACNIDIDÆ.	*Certhiola*	or Yellow-rumped Creeper.
		ALAUDIDÆ.	*Eremophila*	or Skylarks.
		FRINGILLIDÆ or FINCH and SPARROW FAMILY.	*Hesperiphona*	or Evening Grosbeaks.
			Pinicola	" Pine Grosbeaks.
			Carpodacus	" Purple Finches.
			Chrysomitris	" Goldfinches.
			Curvirostra	" Crossbills.
			Ægiothus	" Red Polls.
			Leucosticte	" Gray-crowned Finch.
			Plectrophanes, &c.	" Snow Buntings, &c.
			Passerculus	" Savannah Sparrow, &c.
			Pooecetes	" Grass Finch.
			Coturniculus	" Yellow-winged Sparrow.
			Ammodromus	" Seaside Finch, &c.
			Chondestes	" Lark Finch.
			Zonotrichia	" White-crowned Sparrow.
			Junco	" Snow Birds.
			Poospiza	" Black-throated Sparrow.
			Spizella	" Tree Sparrow, &c.
			Melospiza	" Song Sparrow, &c.
			Peucaca	" Bachman's Finch, &c.
			Embernagra	" Texas Finch.
			Passerella	" Fox-colored Sparrow,
			Calamospiza	" Lark Bunting.
			Euspiza	" Black-throated Bunting.
			Guiraca	" Grosbeaks.
			Cyanospiza, &c.	" Indigo Birds, &c.
			Cardinalis, &c.	" Cardinal Birds.
			Pipilo	" Ground Robins.
		ICTERIDÆ or BLACKBIRD FAMILY.	*Dolichonyx*	or Bobolinks.
			Molothrus	" Cow Birds.
			Agelaius, &c.	" Blackbirds.
			Sturnella	" Larks.
			Icterus	" Orioles.
			Quiscalus	" Grakles.
		STURNIDÆ.	*Sturnus*	or Starlings.
		CORVIDÆ or CROW FAMILY.	*Corvus, &c.*	or Ravens and Crows.
			Pica	" Magpies.
			Cyanura, &c.	" Jays.

[VERTEBRATES: BIRDS — *Continued.*]

ORDERS.		FAMILIES.	Genera.	
RASORES or SCRATCHERS.	COLUMBÆ.	COLUMBIDÆ or DOVE FAMILY.	*Columba*	or Doves.
			Ectopistes	" Wild Pigeon.
			Zenaida	" Zenaida Dove.
			Melopelia	" White-winged Dove.
			Zenaidura	" Carolina Dove.
			Scardafella	" Scaly Dove.
			Chamæpelia	" Ground Dove.
			Oreopeleia	" Key West Dove.
			Starnoenas	" Blue-headed Pigeon.
	GALLINÆ.	GOURIDÆ.	*Goura*	or Crown Pigeons.
		PENELOPIDÆ.	*Ortalida*	or Chiacalacca.
		MEGAPODIDÆ.		or Mound Birds.
		PHASIANIDÆ or PHEASANT FAMILY.	*Meleagris*	or Wild Turkeys.
			Pavo	" Peacocks.
			Numida	" Guinea Fowl.
			Gallus	" Domestic Cock, &c.
			Phasianus	" Pheasants.
		TETRAONIDÆ or GROUSE FAMILY.	*Tetrao*	or Spruce Partridge, &c.
			Centrocercus	" Cock of the Plains.
			Pediocetes	" Sharp-tailed Grouse.
			Cupidonia	" Prairie Chicken.
			Bonasa	" Ruffed Grouse.
			Lagopus	" Ptarmigans.
		PERDICIDÆ or PARTRIDGE FAMILY.	*Ortyx*	or Partridge or Quail.
			Oreortyx	" Plumed Partridge.
			Lophortyx	" California Quail.
			Callipepla	" Blue Partridge.
			Cyrtonyx	" Massena Partridge.
			Perdix	" Europ'n Gray Partridge.
			Coturnix	" European Quail.
CURSORES.		STRUTHIONIDÆ or OSTRICH FAMILY.	*Struthio*	or Ostriches.
			Rhea	" South American Ostrich.
			Casuarius	" Cassowaries.
			Apteryx	
		OTIDÆ.	*Otis*	or Bustards.
GRALLATORES or WADERS.	HERODIONES.	GRUIDÆ.	*Grus*	or Cranes.
		ARAMIDÆ.	*Aramus*	or Courlans.
		ARDEIDÆ or HERON FAMILY.	*Demigretta*	or Egrets.
			Garzetta	" Snowy Herons.
			Herodias	" White Herons.
			Ardea	" Great Blue Herons.
			Audubonia	" Great White Herons.
			Florida	" Blue Herons.
			Ardetta	" Least Bittern.
			Botaurus	" Stake Drivers.
			Butorides	" Green Heron.
			Nyctiardea	" Night Heron.
			Nyctherodius	" Yellow-crowned Heron.
		CANCROMIDÆ.	*Cancroma*	or Boat-Bills.

[VERTEBRATES: BIRDS — *Continued.*]

ORDERS.		FAMILIES.	*Genera.*	
GRALLATORES or WADERS — *Continued.*	HERODIONES — *Continued.*	CINCONIDÆ.	*Cinconia* *Jabiru*	or Storks.
		TANTALIDÆ.	*Tantalus* *Ibis*	or Ibises. " Scarlet Ibis, &c.
		PLATALEIDÆ.	*Platalea*	or Spoonbills.
		PHŒNICOPTERIDÆ.	*Phœnicopterus*	or Flamingo.
	GRALLÆ.	CHARADRIDÆ or PLOVER FAMILY.	*Charadrius* *Ægialitis* *Squatarola*	or Golden Plover. " Kill-deer, &c. " Black-bellied Plover.
		HÆMATOPODIDÆ.	*Hæmatopus* *Strepsilas*	or Oyster-catchers. " Turnstones.
		RECURVIROSTRIDÆ.	*Recurvirostra* *Himantopus*	or Avosets. " Stilts.
		PHALAROPIDÆ.	*Phalaropus*	or Phalaropes.
		SCOLOPACIDÆ or SNIPE FAMILY.	*Philohela* *Gallinago* *Macrorhamphus* *Tringa* *Calidris* *Ereuntes* *Micropalma* *Symphemia* *Glottis* *Gambetta* *Rhyacophilus* *Heteroscelus* *Tringoides* *Philomachus* *Actiturus* *Tryngites* *Limosa* *Numenius*	or Woodcocks. " Snipes. " " " Sandpipers. " Sanderling. " Semi-palmated Sandpip ers. " Stilt Sandpipers. " Willets. " Greenshanks. " Yellow-legs. " Solitary Sandpiper. " Wandering Tatler. " Spotted Sandpiper. " Ruff. " Field Plover. " Buff-breasted Sandpiper. " Godwits. " Curlews.
		RALLIDÆ or RAIL FAMILY.	*Rallus* *Porzana* *Crex* *Fulica* *Gallinula*	or Rails. " Sora Rails. " Corn Crakes. " Coots. " Gallinules.

<table>
<tr><th>ORDERS.</th><th></th><th>FAMILIES.</th><th colspan="2">Genera.</th></tr>
<tr><td rowspan="18">NATATORES or SWIMMERS.</td><td rowspan="2">ANSERES.</td><td rowspan="2">ANATIDÆ
or
DUCK FAMILY.</td><td>Cygnus
Anser
Bernicla, &c.
Anas
Dafila
Nettion
Querquedula
Spatula
Chaulelasmus
Mareca
Aix
Fulix
Aythya</td><td>or Swans.
" White-fronted Goose.
" Wild Goose, &c.
" Mallard, &c.
" Pintail.
" Green-winged Teals,
" Blue-winged Teals, &c.
" Shoveller.
" Gadwall.
" Widgeons.
" Wood Duck.
" Scaup Ducks.
" Canvas-Back and Red-Head.</td></tr>
<tr><td>Bucephala
Histrionicus
Harelda
Camptolæmus
Melanetta
Pelionetta
Oidemia
Somateria, &c.
Erismatura
Mergus
Lophodytes
Mergellus</td><td>" Golden-eye and Dipper.
" Harlequin Duck.
" Longtail.
" Labrador Ducks.
" White-winged Coot.
" Sea Coot.
" Scoter.
" Eider Ducks.
" Ruddy Ducks.
" Sheldrakes.
" Hooded Merganser.
" Smew.</td></tr>
<tr><td rowspan="16">GAVIÆ.</td><td>PELECANIDÆ.</td><td>Pelecanus</td><td>or Pelicans.</td></tr>
<tr><td>SULIDÆ.</td><td>Sula</td><td>or Gannets.</td></tr>
<tr><td>TACHYPETIDÆ.</td><td>Tachypetes</td><td>or Man of-War-Bird.</td></tr>
<tr><td>PHALACROCORACIDÆ.</td><td>Graculus</td><td>or Cormorants.</td></tr>
<tr><td>PLOTIDÆ.</td><td>Plotus</td><td>or Snake-Bird.</td></tr>
<tr><td>PHÆTONIDÆ.</td><td>Phæton</td><td>or Tropic Bird.</td></tr>
<tr><td>PROCELLARIDÆ
or
PETREL FAMILY.</td><td>Diomedea
Procellaria
Thallasidroma
Puffinus</td><td>or Albatrosses.
" Fulmar Petrels.
" Stormy Petrels.
" Shearwaters.</td></tr>
<tr><td>LARIDÆ
or
GULL FAMILY.</td><td>Stercorarius
Larus, &c.
Sterna, &c.
Rhynchops</td><td>or Jagers or Skua Gulls.
" Gulls.
" Terns.
" Black Skimmers.</td></tr>
<tr><td>COLYMBIDÆ.</td><td>Colymbus
Podiceps
Podylimbus</td><td>or Divers.
" Grebes.
" Pied Grebe.</td></tr>
<tr><td>ALCIDÆ
or
AUK FAMILY.</td><td>Alca
Aptenodytes
Mormon
Uria
Mergulus</td><td>or Auks.
" Penguins.
" Puffins.
" Guillemots.
" Sea-Dove or Dove-Kie.</td></tr>
</table>

THE CLASS OF REPTILIA OR REPTILES.

ORDERS.		FAMILIES.	*Genera.*	
TESTUDINATA or TURTLES.	CHELO-AMYDÆ or FRESH-WATER AND LAND TURTLES.	TESTUDININA or LAND TORTOISE FAMILY.	*Xerobates*	or Gophers.
			Testudo	" Galapago Tortoise, and European Tortoise.
		EMYDOIDÆ or TERRAPIN FAMILY.	*Ptychemys*	or Red-bellied Terrapin.
			Trachemys	
			Graptemys	" Map Turtles.
			Malacoclemmys	" Salt-water Terrapins.
			Chrysemys	" Painted Turtles.
			Deirochelys	" Reticulated Turtles.
			Emys	" Blanding's Tortoise.
			Nanemys	" Speckled "
			Calemys	" Mühlenberg's Tortoise.
			Glyptemys	" Wood "
			Actinemys	
			Cistudo	" Box Turtles.
		CINOSTERNOIDÆ.	*Aromochelys* or *Ozotheca*	or Musk Tortoise.
			Thyrosternum	" Cinosternum or Mud Tortoise.
		CHELYDROIDÆ.	*Macroclemmys*	or Alligator Turtle.
			Chelydra	" Snapping "
		HYDRASPIDÆ.	*Hydraspis*	of South America.
		CHELYOIDÆ.	*Chelys*	or Matamata, of S. America.
		TRIONYCHIDÆ.	*Platypeltis*	or Soft-shelled Turtles.
			Amyda and *Aspidonectes*.	
	CHELONII.	CHELONIOIDÆ or LOGGERHEAD FAMILY.	*Chelonia*	or Green Turtles.
			Eretmochelys	" Hawk-bill Turtles.
			Thallassochelys	" Loggerheads.
		SPHARGIDIDÆ.	*Sphargis*	or Leather-backed Turtles.
SAURIA or SAURIANS.	DINOSAURS.		*Iguanodon*	(Fossil.)
			Hylæosaurus	"
			Megalosaurus	"
	CROCODILIANS.		*Alligator*	or Alligators.
			Crocodilus	" Crocodiles.
			Gavialis	" Gavials.
			Cetiosaurus	(Fossil.)
			Teleosaurus	"
	LACERTIANS.		*Thecodontosaurus*, *Palæosaurus*, *Proterosaurus*, and *Mososaurus* (all Fossil).	
			Ameiva	or Striped Lizards.
			Iguana, &c.	" Iguanas, Monitors, Green Lizards, Horned Toads, Geckos, Chameleons, Skinks, &c.
	ENALIOSAURS.		*Ichthyosaurus*	(Fossil.)
			Plesiosaurus	"
	PTEROSAURS.		*Pterodactyl*	(Fossil.)

[VERTEBRATES: REPTILES — *Continued.*]

ORDERS.	FAMILIES.	Genera.	
OPHIDIA or SERPENTS.	BOIDÆ.	*Boa*	or Boas and Anacondas.
		Python	" Pythons.
	COLUBRIDÆ or COLUBER FAMILY.	*Eutænia*	or Striped Snakes.
		Nerodia	" Water "
		Regina	
		Heterodon	" Hog-nose "
		Pituophis	" Pine "
		Scotophis	
		Ophibolus	" Chain Snakes and Chicken Snakes.
		Georgia	" Indigo Snakes.
		Bascanion	" Black "
		Masticophis	" Coach-whip Snakes.
		Leptophis	" Southern Green "
		Chlorosoma	" Northern Green "
		Diadophis	" Ring-necked "
		Rhinostoma	" Scarlet "
		Rhinocheilus	
		Haldea	" Brown "
		Farancia, &c.	" Horn " &c.
	CROTALIDÆ or RATTLESNAKE FAMILY.	*Crotalus*	or Rattlesnakes.
		Crotalophorus	" Prairie Rattlesnakes.
		Ancistrodon	" Copperheads.
		Toxicophis	" Water Moccasins.
	VIPERIDÆ.	*Vipera*	or Vipers of the Old World.
	ELAPIDÆ.	*Elaps*	or Harlequin Snakes.
		Naia	" Cobra, Haje, &c.
	HYDROPHIDÆ.	*Hydrophis*	or Venomous Water Snakes of the Old World.

THE CLASS OF BATRACHIA OR BATRACHIANS.

ORDERS.	FAMILIES.	Genera.	
ANOURA or TAILLESS BATRACHIANS.	RANIDÆ or FROG FAMILY.	*Rana*	or Frogs proper.
	HYLOIDÆ.	*Hyla*	or Tree-Frogs or Tree-Toads.
		Hylodes	" Cricket-Frogs.
	BUFONIDÆ or TOAD FAMILY.	*Bufo*	or Toads.
		Scaphiophus	" Toad-Frog.
URODOLA or TAILED BATRACHIANS.	SALAMANDRIDÆ or SALAMANDER FAMILY.	*Salamandra* and *many other genera*	or Salamanders & Tritons.
	AMPHIUMIDÆ.	*Amphiuma*	or Congo Snake.
		Menopoma	" Hell-bender.
	SIRENIDÆ or SIREN FAMILY.	*Siren*	or Sirens.
		Menobranchus	" Mud-Puppies.
		Siredon	" Axolotl.
		Proteus	
APODA or FOOTLESS BATRACHIANS.	CÆCILIIDÆ.	*Cæcilia*	or Cæcilians, Blind-worms.

THE CLASS OF FISHES.*

ORDERS.		FAMILIES.	*Genera.*	
SELACHI or SELACHIANS or PLAGIOSTOMI.	RAIÆ or RAYS or SKATES.	CEPHALOPTERIDÆ.	*Cephaloptera, &c.*	or Vampires, &c.
		MYLIOBATIDÆ.	*Myliobates, &c.*	or Sea-Eagles, &c.
		TRYGONIDÆ.	*Trygon, &c.*	or Sting Rays, &c.
		RAIIDÆ.	*Raia, &c.*	or Rays or Skates proper.
		TORPEDINIDÆ.	*Torpedo, &c.*	or Electric Rays.
		RHINOBATIDÆ.	*Rhinobatus, &c.*	or Rhinobats.
		PRISTIDÆ.	*Pristis*	or Saw-Fish.
	SQUALI or SHARKS.	ZYGÆNIDÆ.	*Zygæna*	or Hammerhead Shark.
		SQUATINIDÆ.	*Squatina*	or Monk Fish.
		SCYMNIDÆ.	*Scymnus, &c.*	or Nurse or Sleeper, &c.
		SPINACIDÆ.	*Acanthias, &c.*	or Spined Dog-Fish, &c.
		NOTIDANIDÆ.	*Hexanchus, &c.*	
		RHINODONTIDÆ.	*Rhinodon*	
		CESTRACIONTIDÆ.	*Cestracion*	or Cestracionts.
		ALOPECIIDÆ.	*Alopecias*	or Thresher Shark.
		LAMNIDÆ.	*Lamna* *Selachus*	or Mackerel Shark. " Basking or Elephant Shark.
		GALEIDÆ.	*Mustelus, &c.*	or Topes and Hounds.
		CHARCARIDÆ.	*Carcharias*	or White Shark, Gray Shark, &c.
		SCYLLIDÆ.	*Scyllium*	or Dog-Fishes.
		†		
GANOIDEI or GANOIDS.		STURIONIDÆ.	*Accipenser, &c.*	or Sturgeons.
		AMIIDÆ.	*Amia*	or Mud-Fishes.
		POLYPTERIDÆ.	*Polypterus*	or Polypterus of the Nile.
		LEPIDOSTEIDÆ.	*Lepidosteus*	or Gar-Pikes.
TELEOSTEI or BONY FISHES.	LOPHOBRANCHII.	SYGNATHIDÆ.	*Sygnathus, &c.* *Hippocampus, &c.*	or Pipe-Fishes, &c. " Sea-Horses, &c.
		PEGASIDÆ.	*Pegasus*	or Flying-Horses.
	PLECTOGNATHI.	DIODONTIDÆ.	*Diodon* *Tetraodon* *Orthagoriscus*	or Puffers. " " " Sun-Fish.
		OSTRACIONIDÆ.	*Ostracion, &c.*	or Trunk-Fish, &c.
		BALISTIDÆ.	*Balistes, &c.*	or File-Fishes.

* The classification here adopted is essentially that of Müller, as modified by Owen, and is taken from the Encyclopædia Britannica.

† Owen places here Chimæridæ and Sirenidæ, the latter represented by the Lepidosiren.

[VERTEBRATES: FISHES—*Continued.*]

ORDERS.		FAMILIES.	*Genera.*	
TELEOSTEI or BONY FISHES—*Continued.*	ACANTHOPTERI or SPINE-FINNED FISHES.	LOPHIIDÆ.	*Lophius* *Cheironectes* *Batrachus*	or Angler. " Hand Fishes. " Toad "
		BLENNIIDÆ.	*Blennius* *Zoarces* *Anarrichas, &c.*	or Blennies. " Eel-Pout. " Wolf-Fish, &c.
		GOBIIDÆ.	*Gobius, &c.* *Cyclopterus*	or Gobies, &c. " Lump-Fish.
		AULOSTOMIDÆ.	*Fistularia, &c.*	or Flute-mouths.
		TEUTHYDIDÆ.	*Acanthurus, &c.*	or Lancet-Fish, &c.
		TÆNIIDÆ.	*Trachypterus, &c.*	or Ribbon-Fish.
		CHÆTODONTIDÆ.	*Chætodon, &c.*	or Chætodonts.
		ZEIDÆ.	*Zeus, &c.*	or Dories, &c.
		SCOMBRIDÆ or MACKEREL FAMILY.	*Scomber* *Thynnus* *Xiphias* *Pelamys* *Naucrates* *Temnodon* *Coryphæna, &c.*	or Mackerels proper. " Tunnies. " Sword-Fish. " Skip-Jack. " Pilot-Fish. " Blue-Fish. " Dolphins, &c.
		ATHERINIDÆ.	*Atherina*	or Silversides.
		MUGILIDÆ.	*Mugil, &c.*	or Mullets.
		LABYRYNTHIBRANCHIDÆ.	*Anabas, &c.*	or Climbing Perch, &c.
		SPARIDÆ.	*Sargus* *Pagrus, &c.*	or Sheepshead. " Scupaug or Porgee, &c.
		SCIÆNIDÆ.	*Otolithus, &c.* *Corvina* *Umbrina* *Pogonias, &c.*	or Weak-Fish. " Lake Sheepshead. " King-Fish. " Drums, &c.
		TRIGLIDÆ or SCLEROGENIDÆ, &c. or SCULPIN FAMILY, &c.	*Trigla* *Prionotus* *Dactylopterus* *Cottus* *Hemitripterus* *Scorpæna* *Gasterosteus, &c.*	or Gurnards. " Sea-Robins. " Sea-Swallows. " Sculpins. " Sea-Raven. " Sea-Scorpion. " Sticklebacks, &c.
		MULLIDÆ.	*Mullus, &c.*	or Surmullets.
		POLYNEMIDÆ.	*Polynemus, &c.*	or Paradise Fish, &c.
		THERAPONIDÆ.	*Pomotis* *Boleosoma, &c.*	or Breams. " Darters, &c.
		PERCIDÆ.	*Perca* *Labrax, &c.*	or Perch proper. " Bass, &c.
		URANOSCOPIDÆ.	*Uranoscopus, &c.*	or Star-Gazers, &c.

[VERTEBRATES: FISHES — *Continued.*]

ORDERS.		FAMILIES.	*Genera.*	
TELEOSTEI or BONY FISHES — *Continued.*	ANACANTHINI.	PLEURONECTIDÆ.	*Platessa* *Hippoglossus* *Rhombus &* *Solea, &c.*	or Flounders. " Halibuts. " Turbots and Soles, &c.
		ECHENEIDIDÆ.	*Echeneis*	or Remora.
		GADIDÆ or COD FAMILY.	*Morrhua* *Merlangus* *Merluccius* *Lota* *Brosmius* *Phycis, &c.*	or Cods and Haddock. " Pollack. " Whiting. " Burbot. " Cusk. " Hake, &c.
		OPHIDIDÆ.	*Ophidia*	or Serpent-form Fishes.
	PHARYNGO-GNATHI.	AMBIOTOCIDÆ.	*Ambiotoca*	
		LABRIDÆ.	*Ctenolabrus* *Tautoga, &c.*	or Conners. or Tautog, &c.
		SCOMBERESO-CIDÆ.	*Belone* *Scomberesox, &c.* *Exocœtus*	or Gar-Fishes. " Bill-Fishes, &c. " Flying-Fishes.
	MALACOPTERI.	GONIODONTIDÆ.	*Goniodontes*	or Goniodonts.
		SILURIDÆ.	*Silurus, &c.* *Pimelodus, &c.*	or Silurus, &c. " Cat-Fishes.
		CYPRINIDÆ.	*Cyprinus, &c.* *Leuciscus, &c.*	or Carps proper, &c. " Dace, Shiners, &c.
		CATOSTOMIDÆ.	*Catostomus*	or Suckers.
		ESOCIDÆ.	*Esox*	or Pike and Pickerel.
		CYPRINODON-TIDÆ.	*Fundulus &* *Hydrargyra, &c.*	or Mummachogs, &c.
		MORMYRIDÆ.	*Mormyrus.*	
		ELOPIDÆ.	*Elops*	or Silver-Fish.
		CLUPESOCIDÆ.	*Notopterus, &c.*	or Herring-Pikes.
		SCOPELIDÆ.	*Scopelus, &c.*	
		CHARACINIDÆ.	*Salminus, &c.*	or Salmon-like Fishes.
		SALMONIDÆ.	*Salmo, &c.*	or Salmon, Trout, Smelts, &c.
		CLUPEIDÆ.	*Clupea* *Alausa, &c.*	or Herring, Pilchards, &c. " Shad, Alewives, &c.
		APHRODEIRIDÆ.		or L. Ponchartrain Fishes.
		HETEROPYGII.	*Amblyopsis*	or Blind-Fish.
		GYMNOTIDÆ.	*Gymnotus, &c.*	or Electric Eels.
		OPHISURIDÆ.	*Ophisurus, &c*	or Snake-tailed Eels.
		CONGERIDÆ.	*Conger*	or Conger Eels.
		ANGUILLIDÆ.	*Anguilla*	or Common Eels.
		MURÆNIDÆ.	*Muræna, &c.*	or Roman Muræna.
		SYNBRANCHIDÆ.		or Eel-like Fishes.

[VERTEBRATES: FISHES — *Continued.*]

ORDERS.	FAMILIES.	Genera.	
DERMOPTERI.	LEPTOCEPHALIDÆ.	*Leptocephalus*	or Slender-heads.
	PETROMYZONTIDÆ.	*Petromyzon*	or Lampreys.
	MYXINIDÆ.	*Myxine*	or Hag.
	AMMOCŒTIDÆ.	*Ammocœtes*	or Mud-Lampreys.
	AMPHIOXIDÆ.	*Branchiostoma*	or Amphioxus.

THE BRANCH OF ARTICULATA, OR ARTICULATES.

THE CLASS OF INSECTA, OR INSECTS.

INSECTS PROPER.	HYMENOPTERA or BEES, WASPS, &c.	APIARIE or APIDÆ.	*Apis* *Bombus* *Xylocopha, &c.*	or Hive Bee. " Humble-Bees. " Carpenter Bees, &c.
		VESPARIÆ or VESPIDÆ.	*Vespa* *Polistes, &c.*	or Wasps and Hornets. " Wasps.
		CRABRONIDÆ.	*Crabro, &c.*	or Wood-Wasps, &c.
		BEMBECIDÆ.	*Bembex, &c.*	
		SPHEGIDÆ.	*Sphex, &c.*	or Mud Wasps.
		SCOLIETÆ.	*Scolia, &c.*	
		FORMICARIÆ.	*Formica, &c.*	or Ants.
		CHRYSIDIDÆ.	*Chrysis, &c.*	or Golden Wasps.
		PROCTOTRUPIDÆ	*Platygaster, &c.*	or Egg-Parasites.
		CHALCIDIDÆ.	*Chalcis, &c.*	
		ICHNEUMONIDÆ.	*Ichneumon*	or Ichneumons.
		EVANIALES.	*Evania, &c.*	
		CYNIPSERA.	*Cynips, &c.*	or Gall Flies.
		UROCERATA.	*Tremex, &c.*	or Boring-Saw Flies.
		TENTHREDINETÆ.	*Selandria,* *Cimbex, &c.*	or Rose & Elm Saw-Flies, &c.
	LEPIDOPTERA or BUTTERFLIES and MOTHS.	PAPILIONIDÆ.	*Papilio, &c.*	or Papilio Butterflies, &c.
		PIERIDÆ.	*Pieris* *Colias*	or White Butterfly. " Yellow "
		NYMPHALIDÆ.	*Limenitis, Danais, Argynnis, &c.*	
		SATYRIDÆ.	*Satyrus, &c.*	or Hipparchians, &c.
		LYCÆNIDÆ.	*Chrysophanus, &c.*	or Copper Butterflies, &c.
		HESPERIDÆ, &c.	*Hesperia, &c.*	or Skipper Butterflies.
		SPHINGIDÆ.	*Sphinx, &c.*	or Hawk-Moths, &c.
		ÆGERIDÆ.	*Trochilium, &c.*	or Peach-tree Borers, &c.
		ZYGÆNIDÆ.	*Eudryas, &c.*	or Wood Nymphs, &c.
		BOMBYCIDÆ.	*Bombyx, Telea, &c.*	or Silk-Worm Moths.
		NOCTUELITÆ.	*Agrotis, &c.*	or Dart-Moths, &c.
		PHALÆNIDÆ.	*Geometra, &c.*	or Geometers, Canker-worms, &c.
		PYRALIDÆ.	*Pyralis, &c.*	or Meal-Moth, &c.
		TORTRICIDÆ.	*Penthina, &c.*	or Apple-worm Moth, &c.
		TINEIDÆ.	*Tinea, &c.*	or Clothes Moths, &c.
		PTEROPHORII.	*Pterophorus*	or Feather-winged Moths.

[ARTICULATES: INSECTS — *Continued.*]

ORDERS.		Families.	Genera.	
INSECTS PROPER — *Continued.*	DIPTERA or FLIES, &c.	CULICIDÆ	*Culex, &c.*	or Gnats, Mosquitoes, &c.
		TIPULARIÆ.	*Ceidomyia,* *Tipula, &c.*	or Hessian Fly, &c.
		TABANIDÆ.	*Tabanus*	or Horse Flies.
		ASILICI.	*Asilus*	or Asilus Flies.
		BOMBYLIARII.	*Bombylius*	or Bee Flies.
		SYRPHIDÆ.	*Syrphus, &c.*	or Syrphus Flies.
		DOLICHOPIDÆ.	*Dolichopus*	or Long-legged Flies.
		ŒSTRIDÆ.	*Gasterophilus,* *Œstrus, &c.*	or Bot Flies.
		MUSCIDÆ.	*Musca, &c.*	or House Flies, &c.
		HIPPOBOSCIDÆ.	*Hippobosca, &c.*	or Spider Flies.
		PULICIDÆ.	*Pulex*	or Fleas.
	COLEOPTERA or BEETLES.	CICINDELIDÆ.	*Cicindela*	or Tiger Beetles.
		CARABIDÆ.	*Calasoma, &c.*	or Caterpillar Hunters.
		DYTICIDÆ.	*Dyticus*	or Water Beetles.
		GYRINIDÆ.	*Gyrinus*	or Whirligig Beetles.
		HYDROPHILIDÆ.	*Hydrophilus*	or Water-loving Beetles.
		SILPHIDÆ.	*Silpha*	or Carrion Beetles.
		STAPHYLINIDÆ.	*Staphylinus*	or Rove Beetles.
		HISTERIDÆ.	*Hister*	or Mimic Beetles.
		DERMESTIDÆ.	*Dermestes*	or Skin Beetles.
		BYRRHIDÆ.	*Byrrhus*	
		LUCANIDÆ.	*Lucanus*	or Horn-Bugs.
		SCARABÆIDÆ.	*Copris* *Geotrupes* *Macrodactylus* *Lachnosterna, &c.*	or Tumble Beetles. " Earth-borers. " Rose-chafers. " May Beetles, &c.
		BUPRESTIDÆ.	*Buprestis*	or Buprestians.
		ELATERIDÆ.	*Elater*	or Snap or Spring Beetles.
		LAMPYRIDÆ.	*Lampyris*	or Fire-Flies or Glow-worms.
		MALACHIDÆ.		
		CLERIDÆ.	*Clerus*	or Bee-destroyers.
		LYMEXILLIDÆ.	*Lymexylon*	or Wood-destroyers.
		PTINIDÆ.	*Anobius, &c.*	or Death-Watches, &c.
		TENEBRIONIDÆ.	*Tenebrio*	or Meal-worms.
		MORDELLIDÆ.	*Mordella*	
		MELOIDÆ.	*Cantharis, &c.*	or Cantharides.
		STYLOPIDÆ.	*Stylops, &c.*	or Bee-Parasites.
		CURCULIONIDÆ.	*Curculio, &c.*	or Curculios or Weevils.
		CERAMBYCIDÆ.	*Prionus, &c.*	or Capricorn Beetles.
		CHRYSOMELIDÆ.	*Chrysomela, &c.*	or Chrysomelians, &c.
		COCCINELLIDÆ.	*Coccinella*	or Lady-Birds.

[ARTICULATES: INSECTS— *Continued.*]

ORDERS.		FAMILIES.	*Genera.*	
INSECTS PROPER — *Continued.*	HEMIPTERA or CICADAS, BUGS, &c.	CICADARIÆ.	*Cicada*	or Cicadas or Harvest Flies.
		FULGORIDÆ.	*Fulgora*	or Lantern-Flies.
		CERCOPIDÆ.	*Membracis*	or Tree-Hoppers.
		APHIDÆ.	*Aphis*	or Plant-Lice.
		COCCIDÆ.	*Coccus*	or Bark-Lice, Cochineal, &c.
		NOTONECTIDÆ.	*Notonecta*	or Boat Flies.
		NEPIDÆ.	*Nepa*	or Scorpion-Bugs.
		HYDROMETRIDÆ.	*Gerris*	or Water-Measurers.
		COREIDÆ.	*Coreus, &c.*	or Squash-Bug, &c.
		THRIPSIDÆ.	*Thrips*	
		CIMICIDÆ.	*Cimex*	or Bed-Bug.
		PEDICULIDÆ.	*Pediculus, &c.*	or Lice.
	ORTHOPTERA or GRASSHOPPERS, &c.	FORFICULIDÆ.	*Forficula*	or Earwigs.
		BLATTARIÆ.	*Blatta*	or Cockroaches.
		PHASMIDA.	*Diaphomera, &c.*	or Walking-stick, &c.
		MANTIDÆ.	*Mantis*	or Mantis.
		GRYLLIDES.	*Gryllus, &c.*	or Field Crickets, &c.
		LOCUSTARIÆ.	*Cyrtophyllus, &c.*	or Katydid, &c.
		ACRYDII.	*Caloptenus, &c.*	or Red-legged Locust, &c.
		THYSANOURA.	*Lepisma, &c.*	or Spring-tails.
	NEUROPTERA or DRAGON-FLIES, &c.	TERMITIDÆ.	*Termites*	or Termites.
		PSOCIDÆ.	*Atropos*	or Book-Louse, &c.
		PERLARIÆ.	*Perla, &c.*	or Stone-Flies.
		EPHEMERIDÆ.	*Ephemera*	or May-Flies.
		ODONATA.	*Agrion, Æschna, &c.*	or Dragon-Flies.
		SIALIDÆ.	*Corydalis, &c.*	or Corydalis, &c.
		HEMEROBINI.	*Hemerobius, &c.*	or Lace-wings.
		PHRYGANIDÆ.	*Neuronia, &c.*	or Caddice Flies.
ARACHNIDA	or SPIDERS, &c.	ARANEIDÆ.	*Lycosa, &c.*	or True Spiders.
		PEDIPALPI.	*Buthus, &c.*	or Scorpions.
		PSEUDO-SCORPIONES.	*Chelifer, &c.*	or Book Spiders, &c.
		PHALANGITA.		or Daddy-long-legs.
		ACARINA.	*Trombidium, &c.*	or Velvet-red Mites, &c.
MYRIAPODA	or CENTIPEDES.	GLOMERIDÆ.	*Glomeris*	
		JULIDÆ.	*Julus*	or Galley-Worm.
		POLYDESMIDÆ.	*Polydesmus.*	
		LITHOBIIDÆ.	*Lithobius*	or Earwigs.
		SCOLOPENDRIDÆ.	*Scolopendra*	or Centepedes.
		GEOPHILIDÆ.	*Geophilus*	

THE CLASS OF CRUSTACEA OR CRUSTACEANS.

ORDERS.		FAMILIES.	*Genera.*	
DECAPODA or TEN-FOOTED CRUSTACEANS.	BRACHYURANS (including ANOMURANS) or CRABS.	MAIIDÆ or SEA-SPIDER FAMILY.*	*Maia, &c.*	or Sea-Spiders, &c.
		CANCRIDÆ or EDIBLE CRAB FAMILY.	*Cancer, &c.*	or Edible Crab of Europe, &c.
		PORTUNIDÆ or EDIBLE CRAB FAMILY.	*Lupa, &c.*	or Edible Crab of U. S., &c.
		GECARCINIDÆ or LAND CRAB FAMILY.	*Gecarcinus, &c.*	or Land Crabs.
		GELASMIDÆ or FIDDLER CRAB FAMILY.	*Gelasmus, &c.*	or Fiddler Crabs.
		PAGURIDÆ or HERMIT CRAB FAMILY.	*Pagurus, &c.*	or Hermit Crabs.
	MACRURANS or LOBSTERS, &c.	PALINURIDÆ or SPINY LOBSTER FAMILY.	*Palinurus, &c.*	or Spiny Lobsters.
		ASTICIDÆ or COMMON LOBSTER FAMILY.	*Homarus* *Astacus, &c.*	or Common Lobsters. " Cray-Fishes.
	GASTRURANS or STOMAPODS or SHRIMPS, &c.	CRANGONIDÆ or SHRIMP FAMILY.	*Crangon, &c.*	or Shrimps.
		PALEMONIDÆ or PRAWN FAMILY.	*Palemon, &c.*	or Prawns.
		SQUILLIDÆ or SEA MANTIS FAMILY.	*Squilla, &c.*	or Sea Mantes.
		MYSIDÆ or OPOSSUM SHRIMP FAMILY.	*Mysis, &c.*	or Opossum Shrimps.

* Only the more common Families and Genera of Crustaceans are here given. The same is true in regard to Worms, on page XXI.

[ARTICULATES: CRUSTACEANS — *Continued.*]

ORDERS.		FAMILIES.	*Genera.*	
TETRADECAPODA or FOURTEEN-FOOTED CRUSTACEANS.	ISOPODS (including ANISOPODS).	ONISCIDÆ or WOOD LOUSE FAMILY.	*Oniscus*	or Wood Louse.
		ARMADILLIDÆ or PILL-BUG FAMILY.	*Armadillo*	or Pill Bugs.
		CYMOTHOIDÆ.	*Cymathoa*	or Parasites.
		BOPYRIDÆ.	*Bopyrus*	or Parasites.
	AMPHIPODS (including LÆMIDOPODS).	ORCHESTIDÆ or SAND-HOPPER FAMILY.	*Orchestia*	or Sand-Hoppers or Beach-Fleas.
		GAMMARIDÆ or FRESH-WATER SHRIMP FAMILY.	*Gemmarus*	or Fresh-water Shrimps.
		CAPRELLIDÆ or MEASURER FAMILY.	*Caprella*	or Measurers.
		CYAMIDÆ or WHALE-LOUSE FAMILY.	*Cyamus*	or Whale-Lice.
	TRILOBITES. (Fossil.)			
ENTOMOSTRACA or ENTOMOSTRACANS.	CARCNOIDS (including SIPHONOSTOMES.)	CYCLOPIDÆ or CYCLOPS FAMILY.	*Cyclops*	
		ARGULIDÆ.	*Argulus*	
		CALIGIDÆ.	*Caligus*	
	OSTRACOIDS (including CIRRIPEDS).	CYPRIDÆ or CYPRIS FAMILY.	*Cypris*	
		DAPHNIADÆ or MONOCULUS FAMILY.	*Daphnia*	or Monoculus.
		LIMNADIADÆ.	*Limnadia*	
		LEPADIDÆ.	*Anatifa*	or Geese Barnacles.
		BALANIDÆ.	*Balanus*	or Acorn Barnacles.
	LIMULOIDS.	LIMULIDÆ.	*Limulus*	or Horse-shoe Crab.
	ROTIFERA.			

THE CLASS OF WORMS.

ORDERS.		Families.	Genera.	
ANNELIDES.	DORSIBRANCHIATÆ.	ARENICOLIDÆ or SAND-WORM FAMILY	*Arenicola*	or Sand-Worms or Lob-Worms.
	TUBICOLÆ.	SERPULIDÆ or SERPULA FAMILY.	*Serpula*	or Serpulæ.
	TERRICOLÆ.	LUMBRICIDÆ or EARTH-WORM FAMILY.	*Lumbricus*	or Earth-Worms.
TREMATODS or SUCTORIA.		HIRUNDINIDÆ or LEECH FAMILY.		
NEMATOIDS.	HELMINTHES or ENTOZOA.	CESTOIDÆ or TAPE-WORM FAMILY.	*Tænia*	or Tape-Worms.
		GORDIIDÆ or HAIR-WORM FAMILY.	*Gordius*	or Hair-Worms.

THE BRANCH OF MOLLUSCA OR MOLLUSKS.

THE CLASS OF CEPHALOPODA OR CEPHALOPODS.

ORDERS.	Families.	Genera.	
DIBRANCHIATA or TWO-GILLED CEPHALOPODS.	ARGONAUTIDÆ or PAPER SAILOR FAMILY.	*Argonauta*	or Argonaut or Paper Sailor.
	OCTOPODIDÆ or POULPE FAMILY.	*Octopus*	or Poulpes.
	TEUTHIDÆ or SQUID FAMILY.	*Loligo*	or Squids.
	BELEMNITIDÆ or BELEMNITE FAMILY. (Fossil.)		
	SEPIADÆ or CUTTLE-FISH FAMILY.	*Sepia*	or Cuttle-Fishes.
	SPIRULIDÆ.	*Spirula*	or Spirulas.
TETRABRANCHIATA or FOUR-GILLED CEPHALOPODS.	NAUTILIDÆ or NAUTILUS FAMILY.	*Nautilus*	or Pearly Nautilus.
	ORTHOCERATIDÆ or ORTHOCERAS FAMILY. (Fossil.)		
	AMMONITIDÆ or AMMONITE FAMILY. (Fossil.)		

THE CLASS OF GASTEROPODA OR GASTEROPODS.

ORDERS.	FAMILIES.	Genera.	
GASTEROPODA or GASTEROPODS PROPER.	STROMBIDÆ or STROMB FAMILY.	*Strombus* *Pteroceras, &c.*	or Shombs. " Scorpion Shells, &c.
	MURICIDÆ or MUREX FAMILY.	*Murex* *Pyrula, &c.*	or Thorny Woodcock. " Pyrulas, &c.
	BUCCINIDÆ or WHELK FAMILY.	*Buccinum* *Harpa, &c.*	or Whelks. " Harp-Shells, &c.
	CONIDÆ or CONE FAMILY.	*Conus, &c.*	or Cones, &c.
	VOLUTIDÆ or VOLUTE FAMILY.	*Voluta* *Mitra, &c.*	or Volutes. " Mitre-Shells, &c.
	CYPRÆIDÆ or COWRY FAMILY.	*Cypræa* *Ovulum, &c.*	or Cowries. " Egg-Cowries, &c.
	NATICIDÆ or NATICA FAMILY.	*Natica* *Sigaretus, &c.*	or Naticas.
	PYRAMIDELLIDÆ or PYRAMID SHELL FAMILY.	*Pyramidella, &c.*	or Pyramid Shells, &c.
	CERITHIADÆ or CERITHIUM FAMILY.	*Cerithium, &c.*	or Cerithiums.
	MELANIADÆ or MELANIA FAMILY.	*Melania, &c.*	or Melanias.
	TURRITELLIDÆ or WENTLE-TRAP FAMILY.	*Turritella* *Scalaria, &c.*	or Tower-Shells. " Wentle-traps, &c.
	LITORINIDÆ or PERIWINKLE FAMILY.	*Litorina, &c.*	or Periwinkles, &c.
	PALUDINIDÆ or RIVER-SNAIL FAMILY.	*Paludina, &c.*	or River Snails, &c.
	NERITIDÆ or NERITE FAMILY.	*Nerita, &c.*	or Nerites, &c.
	TURBINIDÆ or TOP-SHELL FAMILY.	*Trochus, &c.*	or Top-Shells, &c.
	HALIOTIDÆ or EAR-SHELL FAMILY.	*Haliotis*	or Ear-Shells.

[MOLLUSKS: GASTEROPODS — *Continued*.]

ORDERS.	FAMILIES.	GENERA.	
GASTEROPODS PROPER — *Continued.*	JANTHINIDÆ or VIOLET-SNAIL FAMILY.	*Janthina*	or Violet Snails.
	FISSURELLIDÆ or KEY-HOLE LIMPET FAMILY.	*Fissurella, &c.*	or Key-hole Limpets.
	CAYLYPTRÆIDÆ or BONNET LIMPET FAMILY.	*Calyptræa, &c.*	or Bonnet Limpets.
	PATELLIDÆ or LIMPET FAMILY.	*Patella, &c.*	or Limpets.
	DENTALIDÆ or TOOTH-SHELL FAMILY.	*Dentalium*	or Tooth-Shells.
	CHITONIDÆ or CHITON FAMILY.	*Chiton*	or Chitons.
	HELICIDÆ or LAND-SNAIL FAMILY.	*Helix, &c.*	or Land-Snails.
	LIMACIDÆ or SLUG FAMILY.	*Limax, &c.*	or Slugs.
	LIMNÆIDÆ or POND-SNAIL FAMILY.	*Limnæa* *Physa* *Planorbis, &c.*	
	AURICULIDÆ.	*Auricula, &c.*	or Little-Ear Shells.
	CYCLOSTOMIDÆ.	*Cyclostoma, &c.*	or Round-Mouths.
	ACICULIDÆ.	*Acicula, &c.*	or Needle-Shells.
	TORNATELLIDÆ.	*Tornatella, &c.*	
	BULLIDÆ.	*Bulla, &c.*	or Bubble-Shells.
	APLYSIADÆ.	*Aplysia, &c.*	or Sea-Hares.
	DORIDÆ.	*Doris, &c.*	or Sea-Lemons.
	TRITONIADÆ.	*Tritonia, &c.*	
	ÆOLIDÆ.	*Æolis, &c.*	
	ELYSIADÆ.	*Elysia, &c.*	
HETEROPODA or HETEROPODS.	FIROLIDÆ.	*Firola, &c.*	
	ATLANTIDÆ.	*Atlanta, &c.*	
PTEROPODA or PTEROPODS.	HYALEIDÆ.	*Hyalea, &c.*	
	LIMACINIDÆ.	*Limacina, &c.*	
	CLIIDÆ.	*Clio, &c.*	

THE CLASS OF ACEPHALA OR ACEPHALS.

ORDERS.	FAMILIES.	Genera.	
LAMELLIBRANCHIATA or LAMELLIBRANCHIATES.	OSTREIDÆ or OYSTER FAMILY.	*Ostrea, &c.* *Pecten, &c.*	or Oysters. " Pectens or Scallops.
	AVICULIDÆ or PEARL-OYSTER FAMILY.	*Avicula, &c.*	or Pearl-Oysters, &c.
	MYTILIDÆ or SEA-MUSSEL FAMILY.	*Mytilus, &c.*	or Sea-Mussels.
	ARCADÆ.	*Arca* *Leda, &c.*	
	TRIGONIDÆ or TRIGONIA FAMILY.	*Trigonia, &c.*	
	UNIONIDÆ or POND & RIVER MUSSEL FAMILY.	*Unio* *Anodon, &c.*	
	CHAMIDÆ.	*Chama, &c.*	
	TRIDACNIDÆ or TRIDACNA FAMILY.	*Tridacna* *Hippopus*	
	CARDIADÆ or COCKLE FAMILY.	*Cardia, &c.*	
	LUCINIDÆ.	*Lucina, &c.*	
	CYCLADIDÆ.	*Cyclas, &c.*	
	CYPRINIDÆ.	*Cyprina* *Astarte, &c.*	
	VENERIDÆ or VENUS-SHELL FAMILY.	*Venus* *Cytherea, &c.*	

[MOLLUSKS: ACEPHALS — *Continued.*]

ORDERS.	FAMILIES	*Genera.*	
LAMELLIBRANCHIATES — *Continued.*	MACTRIDÆ.	*Mactara, &c.*	
	TELLINIDÆ.	*Tellina, &c.*	
	SOLENIDÆ or RAZOR-SHELL FAMILY.	*Solen, &c.*	or Razor-Shells, &c.
	MYACIDÆ or CLAM FAMILY.	*Mya, &c.*	or Clams, &c.
	ANATINIDÆ or LANTERN-SHELL FAMILY.	*Pandora, &c.*	
	GASTROCHÆNIDÆ or WATERING-POT-SHELL FAMILY.	*Aspergillum, &c.*	or Watering-pot Shells, &c.
	PHOLADIDÆ or SHIP-WORM FAMILY.	*Pholas* *Teredo*	or Pholads. " Ship-Worms.

THE CLASS OF BRACHIOPODA OR BRACHIOPODS.

	TERREBRATULIDÆ.	*Terebratula, &c.*	
	RHYNCHONELLIDÆ.	*Rhynchonella.*	
	CRANIADÆ.	*Crania.*	
	DISCINIDÆ.	*Discina.*	
	LINGULIDÆ.	*Lingula.*	

THE CLASS OF TUNICATA OR TUNICATES.

	ASCIDIADÆ.	*Ascidium*	or Simple Ascidians.
	CLAVELLINIDÆ.	*Clavellina, &c.*	or Social Ascidians.
	BOTRYLLIDÆ.	*Botryllus, &c.*	or Compound Ascidians.
	PYROSOMIDÆ.	*Pyrosoma*	or Fire-bodies.
	SALPIDÆ.	*Salpa*	or Salps.

THE CLASS OF POLYZOA OR POLYZOANS.

THE BRANCH OF RADIATA OR RADIATES.

THE CLASS OF ECHINODERMATA OR ECHINODERMS.

ORDERS.	FAMILIES.	*Genera.*
HOLOTHURIOIDS or SEA-CUCUMBERS.		
ECHINOIDS or SEA-URCHINS.		
ASTERIOIDS or STAR-FISHES.		
OPHIURIOIDS or SERPENT STARS.		
CRINOIDS or LILY-LIKE ECHINO-DERMS.		

THE CLASS OF ACALEPHA OR JELLY-FISHES.*

ORDERS.	FAMILIES.	*Genera.*
CTENOPHORÆ or COMB-BEARERS.	BOLINIDÆ, &c.	*Bolina, &c.*
	OCYROEÆ.	*Ocyroe, &c.*
	MERTENSIDÆ.	*Mertensia, &c.*
	CYDIPPIDÆ.	*Pleurobrachia, &c.*
	BEROIDÆ.	*Beröe, &c.*
DISCOPHORÆ or DISK JELLY-FISHES.	RHIZOSTOMIDÆ.	*Rhizostoma, &c.*
	POLYCLONIDÆ.	*Polyclonia, &c.*
	AURELIADÆ.	*Aurelia, &c.*
	STHENONIÆ.	*Sthenio, &c.*
	CYANEIDÆ.	*Cyanea, &c.*
	PELAGIDÆ.	*Pelagia, &c.*
	THALLASANTHEÆ.	*Foveolia, &c.*
	TRACHYNEMIDÆ.	*Trachynema, &c.*
	LEUCKARTIDÆ.	*Liriope, &c.*
	CLEISTOCARPIDÆ.	*Manania, &c.*
	ELEUTHEROCARPIDÆ.	*Lucernaria, &c.*
HYDROIDÆ or HYDROIDS.	OCEANIDÆ.	*Oceania, &c.*
	EUCOPIDÆ.	*Eucope, &c.*
	ÆQUORIDÆ.	*Rhegmatodes, &c.*
	GERYONOPSIDÆ.	*Tima, &c.*
	POLYORCHIDÆ.	*Polyorchis, &c.*
	LAODICEIDÆ.	*Lafœa, &c.*
	MELICERTIDÆ	*Melicertum, &c.*
	PLUMULARIDÆ.	*Plumularia, &c.*
	SERTULARIDÆ.	*Sertularia, &c.*
	NEMOPSIDÆ.	*Nemopsis, &c.*
	BOUGAINVILLEÆ.	*Bougainvillia, &c.*
	NUCLEIFERÆ.	*Turris, &c.*
	WILLIADÆ.	*Willia, &c.*
	SARSIADÆ.	*Coryne, &c.*
	ORTHOCORYNIDÆ.	*Corynitis, &c.*
	PENNARIDÆ.	*Pennaria, &c.*
	TUBULARIDÆ.	*Tubularia, &c.*
	HYDRAIDÆ.	*Hydra, &c.*
	HYDRACTINIDÆ.	*Hydractinia, &c.*
	DIPHYIDÆ.	*Eudoxia, &c.*
	AGALMIDÆ.	*Nanomia, &c.*
	PHYSALIDÆ.	*Physalia, &c.*
	VELELLIDÆ.	*Velella, &c.*
	PORPITIDÆ.	*Porpita, &c.*
	MILLEPORIDÆ.	*Millepora, &c.*

* According to Agassiz.

THE CLASS OF POLYPI OR POLYPS.*

ORDERS.	FAMILIES.	*Genera.*
ALCYONARIA or ALCYONARIANS.	PENNATULIDÆ or SEA-PEN FAMILY.	*Pennatula, &c.*
	PAVONARIDÆ.	*Pavonaria, &c.*
	VERETILLIDÆ.	*Veretillum, &c.*
	RENILLIDÆ.	*Renilla, &c.*
	GORGONIDÆ or SEA-FAN FAMILY.	*Gorgonia, &c.*
	PLEXAURIDÆ.	*Muricea, &c.*
	PRIMNOIDÆ.	*Primnoa, &c.*
	GORGONELLIDÆ	*Verrucella, &c.*
	ISIDÆ.	*Isis, &c.*
	CORALLIDÆ or RED CORAL FAMILY.	*Corallium, &c.*
	BRIARIDÆ.	
	ALCYONIDÆ.	*Alcyonium, &c.*
	XENIDÆ.	
	CORNULARIDÆ.	
	TUBIPORIDÆ or ORGAN-PIPE CORAL FAMILY.	*Tubipora, &c.*
ACTINARIA or ACTINARIANS.	ACTINIDÆ or SEA-ANEMONE FAMILY.	*Metridium, &c.*
	THALLASSIANTHIDÆ.	
	MINYIDÆ.	
	ILLYANTHIDÆ.	
	CERIANTHIDÆ.	
	ANTIPATHIDÆ.	
	GERARDIDÆ.	
	ZOANTHIDÆ.	
	BERGIDÆ.	

* According to Verrill.

[RADIATES: POLYPS— *Continued.*]

ORDERS.	FAMILIES.	*Genera.*
MADREPORARIA or MADREPORARIANS.	EUPSAMMIDÆ.	*Astroides, &c.*
	GEMMIPORIDÆ.	
	PORITIDÆ or PORITES FAMILY.	*Porites, &c.*
	MADREPORIDÆ or MADREPORE FAMILY.	*Madrepore, &c.*
	LITHOPHYLLIDÆ	
	MÆANDRINADÆ or MEANDRINA FAMILY.	*Mæandrina, &c.*
	EUSMILLIDÆ.	
	CARYOPHYLL-IDÆ.	*Caryophyllia, &c.*
	STYLINIDÆ.	
	ASTRÆIDÆ or STAR CORAL FAMILY.	*Astræa, &c.*
	OCULINIDÆ or OCULINA FAMILY.	*Oculina, &c.*
	STYLOPHORIDÆ.	
	CYCLOCLITIDÆ.	
	LOPHOSERIDÆ.	
	FUNGIDÆ or FUNGUS CORAL FAMILY.	*Fungia, &c.*
	MERULINIDÆ.	*Merulina, &c.*

THE BRANCH OF PROTOZOA OR PROTOZOANS.

THE END.

www.ingramcontent.com/pod-product-compliance
Lightning Source LLC
LaVergne TN
LVHW020238110826
845151LV00003B/966